Student Study Guide for

Biology

LIFE ON EARTH
and
LIFE ON EARTH WITH PHYSIOLOGY

Eighth Edition

Teresa Audesirk
University of Colorado at Denver and Health Science Center

Gerald Audesirk
University of Colorado at Denver and Health Science Center

Bruce E. Byers
University of Massachusetts, Amherst

PEARSON

Benjamin
Cummings

San Francisco   Boston   New York
Cape Town   Hong Kong   London   Madrid   Mexico City
Montreal   Munich   Paris   Singapore   Sydney   Tokyo   Toronto

W9-BRM-464

Editor-in-Chief: Beth Wilbur
Senior Acquisitions Editor: Chalon Bridges
Executive Director of Development: Deborah Gale
Associate Editor: Leata Holloway
Assistant Editor: Rebecca Johnson
Executive Marketing Manager: Lauren Harp
Executive Managing Editor: Erin Gregg
Managing Editor: Mike Early
Production Supervisor: Shannon Tozier
Project Management: Michael Krapovicky/Pine Tree Composition, Inc.
Composition: Laserwords Private Limited
Cover Design: Maureen Eide and John Christiana
Cover Production: Seventeenth Street Studios
Manufacturing Manager: Pam Augspurger
Text and Cover Printer: Bind-Rite Graphics

Cover Image: Rockhopper Penguins; The Neck, Saunders Island, Falkland Islands, by Laura Crawford
Williams

ISBN 10-digit: 0-13-195769-4
       13-digit: 978-0-13-195769-5

1 2 3 4 5 6 7 8 9 10—BR—11 10 09 08 07
www. aw-bc.com

Table of Contents

To the Student

Welcome to the *Student Study Guide* to accompany *Biology, Life on Earth* Eighth Edition, by Teresa Audesirk, Gerald Audesirk, and Bruce Byers. This supplement has been constructed to help you review the major biological concepts that you will encounter during this course and ask you to think about them critically.

The *Student Study Guide* is divided into forty-four chapters that correspond directly to those in the textbook.

Within each chapter you will find:

- **Chapter Outline**- an outline of the major concepts in the chapter.
- **Learning Objectives**- a point-by-point explanation of the concepts you should understand after reading the chapter.
- **Quizzes**- multiple choice questions that test your understanding of the chapter's major concepts.
- **Answer Key**- answers to the Quizzes that allow you to gauge your understanding of a particular concept or section.

Not only will the *Student Study Guide* help you review important concepts, but it will also provide an opportunity to explore the biology in your life and the world around you. Please visit http://www. aw-bc. com/ audesirk for more study materials and media activities.

CHAPTER 1: AN INTRODUCTION TO LIFE ON EARTH

OUTLINE

Section 1.1 How Do Scientists Study Life?

- Matter has different levels of organization, ranging from the **atom** (the lowest level) to the **community** (the highest level, **Figure 1-1**). The **cell** is the smallest unit of life (**Figure 1-2**).
- Scientific inquiry is based on the principles of (1) **natural causality**, (2) **natural laws that are uniform in space and time**, and (3) the **common perception of natural events**.
- Scientific inquiry occurs by use of the **scientific method**, which consists of **observation**, **question**, **hypothesis**, **prediction**, **experiment**, and **conclusion** (**Figure 1-4**). **Controls** are used to isolate the effect of a single **variable** on an experimental observation.
- Experimental results are useful only if they are communicated to the general public.
- A **scientific theory** is a broad explanation of a natural phenomenon, supported by extensive and reproducible observations. A scientific theory is formed through **inductive reasoning**, which can then be used to support **deductive reasoning**. All scientific theories have the potential to be **disproved**.

Section 1.2 Evolution: The Unifying Theory of Biology

- **Evolution** accounts for both the diversity and similarities of life. Evolution states that modern organisms descended, with modifications, from preexisting life forms.
- Evolution occurs through the processes of (1) **genetic variation** among population members, (2) offspring **inheritance** of genetic variations from parents, and (3) **natural selection**.
- Natural selection occurs when organisms with favorable **adaptations** to their environment survive and pass on these favorable genes to their offspring. Thus, favorable genes are preserved within a population.

Section 1.3 What Are the Characteristics of Living Things?

- All living things exhibit seven characteristics. They (1) are structurally complex and composed of **cells** (**Figure 1-8**), (2) respond to environmental stimuli, (3) maintain internal **homeostasis** (**Figure 1-9**), (4) use acquired **nutrients** to make **energy** (**Figure 1-10**), (5) grow, (6) reproduce, and (7) can evolve.

Section 1.4 How Do Scientists Categorize the Diversity of Life?

- All living things can be grouped into one of three domains: **Bacteria**, **Archaea**, and **Eukarya**. The Eukarya can be further subdivided into three **kingdoms** (**Fungi**, **Plantae**, and **Animalia**) as well as the unicellular **protists** (**Figure 1–11**, **Table 1–1**). Kingdoms are further subdivided onto phyla, classes, orders, families, genera, and species. Species are named using the **binomial system**.
- Bacteria and Archaea lack a nucleus (i.e., are **prokaryotic**), lack membrane-bound organelles, and are typically unicellular. Eukarya have a nucleus (i.e., are eukaryotic) and can be unicellular or multicellular.
- Photosynthetic organisms can make their own energy (i.e., are **autotrophic**), while those that cannot are **heterotrophs**.

Section 1.5 How Does Knowledge of Biology Illuminate Everyday Life?

- The study of biology does not remove the sense of awe that results from living in the world but, in fact, adds to it.

LEARNING OBJECTIVES

Section 1.1 How Do Scientists Study Life?

- Explain the relationship between atoms and molecules.
- Explain the unique properties of complex organic molecules.
- Describe the organizational levels of life from the cell to complex organisms.
- Describe the relationship among species, populations, and communities.
- Discuss the relevance of natural causality in biological study.
- Explain the key to the natural law governing biological study.
- Explain the scientific method from observation through conclusion.
- Describe the relationship between variables and controls in experimentation.
- Discuss the difference between a conclusion and a theory.
- Discuss the differences between a scientific theory and a belief based upon faith.

Section 1.2 Evolution: The Unifying Theory of Biology

- Discuss how genetic variation, inheritance, and natural selection combine to support the theory of evolution.
- Describe adaptations and their role in natural selection.

Section 1.3 What Are the Characteristics of Living Things?

- Discuss the seven characteristics of life.
- Explain the difference between an autotroph and a heterotroph.

Section 1.4 How Do Scientists Categorize the Diversity of Life?

- Describe the differences between prokaryotic and eukaryotic cells.
- Explain the binomial system for naming organisms.

Section 1.5 How Does Knowledge of Biology Illuminate Everyday Life?

- Discuss how the study of biology enhances our appreciation of life around us.

QUIZ 1

1. Science assumes that natural laws (such as the law of gravity) _____.
   - **a.** apply uniformly through space and through time
   - **b.** apply in the laboratory but not in nature
   - **c.** change with time
   - **d.** differ depending on the location of the observer

2. Your baby starts crying. Since she hasn't eaten in two hours, you declare, "The baby must be hungry." Your statement is a(n) _____.
   - **a.** hypothesis
   - **b.** experiment
   - **c.** conclusion
   - **d.** observation

3. A control is needed in an experiment to
   - **a.** provide a comparison with the experimental results obtained when changing the variable.
   - **b.** keep scientists from pursuing unethical questions and practices.
   - **c.** increase the complexity of the experiment.
   - **d.** duplicate the results of the experiment.

4. In a famous experiment that helped disprove the possibility of spontaneous generation, Francesco Redi hypothesized that contrary to popular belief, maggots did not appear spontaneously but came from flies. He predicted that if flies were kept away from meat, then maggots would not appear. To test his hypothesis, Redi placed identical pieces of meat in two identical jars and placed the two jars on the same windowsill. He placed a lid on one jar and left the other uncovered for several days. In the experiment the control is the
   - **a.** jar.
   - **b.** meat in the jar.
   - **c.** time left on the windowsill.
   - **d.** meat in the jar with the lid.

5. Experiments are carefully designed in an attempt to _____.
   - **a.** test a single experimental variable
   - **b.** test multiple experimental variables
   - **c.** prove a hypothesis
   - **d.** none of the above

6. The three natural processes that form the basis for evolution are
   a. adaptation, natural selection, and inheritance.
   b. predation, genetic variation, and natural selection.
   c. mutation, genetic variation, and adaptation.
   d. fossils, natural selection, and adaptation.
   e. genetic variation, inheritance, and natural selection.

7. The diversity of life is mainly due to _____.
   a. atoms
   b. genetic variation
   c. prokaryotic cells
   d. organ systems

8. Which statement is TRUE?
   a. Eukaryotic cells are simpler than prokaryotic cells.
   b. Heterotroph means "self-feeder."
   c. Mutations are accidental changes in genes.
   d. A scientific theory is similar to an educated guess.
   e. Genes are proteins that produce DNA.

9. Your textbook lists seven characteristics that living organisms possess as a group. However, if you could distill these seven characteristics down to two general descriptive properties exclusive to all living organisms, what would they be?
   a. macromolecular complexity and multiple levels of organization
   b. atoms and elements
   c. cellular structure and RNA-based inheritance
   d. chemical relationships and diversity

10. The organic complexity and organization of living organisms depends on the periodic capture of raw materials and energy. Ultimately, the source of these materials and energy is _____.
   a. metabolism
   b. photosynthesis
   c. the sun
   d. other life-forms

11. Energy, such as gasoline for your car, is required for organisms to survive, and even thrive, in the face of diverse environments. An autotrophic organism would _____.
   a. be one that derived its energy from internalizing the cellular matter of other organisms (i.e., eats others)
   b. be one that derived its energy from a renewable external energy source such as sunlight (i.e., photosynthetic organisms)
   c. include cucumbers growing in your garden
   d. includes both (b) and (c)

12. Living systems need energy to _____.
   a. replace worn-out parts and build new parts
   b. help maintain a constant internal environment
   c. grow larger
   d. all of the above

13. You are a NASA scientist and have discovered an organism in outer space that contains its genetic material in the cytoplasm rather than in a nucleus. Given this characteristic, how might you classify this organism?
   a. heterotroph
   b. eukaryote
   c. prokaryote
   d. autotroph

14. A scientist examines an organism and finds that it is eukaryotic, heterotrophic, and multicellular, and it absorbs nutrients. She concludes that the organism is most likely a member of the kingdom _____.
   a. Bacteria
   b. Protista
   c. Plantae
   d. Fungi
   e. Animalia

15. An understanding of basic biological concepts _____.
   a. permits a deeper, and sometimes profound, appreciation of the world around us
   b. provides just a set of facts and ideas about the world around us
   c. is necessary only for biology majors
   d. often decreases our appreciation (i.e., effectively dehumanizes) of the world around us

QUIZ 2

1. Which of the following experiments would best test the following hypothesis: "Bacteria do not arise spontaneously on meat; they are carried onto meat by droplets of water."
   a. Place a piece of meat in a container full of water and determine if bacteria grow.
   b. Place meat in two separate containers: Add droplets of water to one container and no water to the second container. Cover the containers and compare the growth of bacteria in the two containers.
   c. Place meat in two separate containers: Leave one container uncovered and cover the other. Compare the growth of bacteria in the two containers.
   d. Place meat in two flasks: Leave one flask open to the air and bend the neck of the second flask. Compare the growth of bacteria in the two flasks.

3

2. A group of researchers developed a hypothesis and tested it by designing an experiment. The results of the experiment did not support the original hypothesis. What should the researchers do next?
   a. Continue developing experiments to test the hypothesis
   b. Throw out the results that did not support their hypothesis
   c. Reject the original hypothesis and develop a new hypothesis
   d. Conduct an experiment without a control

3. A scientist is testing if a new cancer drug will work on prostate cancer in humans. To properly interpret results of an experiment, control experiments are needed. Which of the following treatments would serve as a control in these experiments, so that the researchers can conclude that the changes that they see in the patients are due to the drug?
   a. Administer different doses of the drug to patients.
   b. Give some patients a "mock" injection that contains the same solution base and the same volume as the drug but does not actually contain the drug itself.
   c. Treat at least 1000 individuals with the drug.
   d. Treat patients at three different hospitals, each in a different state, with the drug.
   e. Both (b) and (d) are correct.

4. In a famous experiment that helped disprove the possibility of spontaneous generation, Francesco Redi hypothesized that contrary to popular belief, maggots did not appear spontaneously but came from flies. He predicted that if flies were kept away from meat, then maggots would not appear. To test his hypothesis, Redi placed identical pieces of meat in two identical jars and placed the two jars on the same windowsill. He placed a lid on one jar and left the other uncovered for several days. In the experiment, the variable being tested is _____.
   a. flies with access to the meat
   b. time exposed to the sun
   c. heat applied to the meat
   d. effect of jar shape on the meat

5. Simple experiments generally isolate and test one _____ **at a time**.
   a. hypothesis
   b. observation
   c. control
   d. variable

6. The process of evolution involves _____.
   a. natural selection of organisms that produce more offspring under certain environmental conditions
   b. the use of the scientific method in the laboratory
   c. an individual organism acquiring a trait in response to a change in its environment
   d. all of the above are correct

7. Natural selection would be best illustrated by which of the following?
   a. a bacterial cell in the human body that dies when a person takes antibiotics
   b. a bacterial cell in the human body with a genetic variation that allows it to survive when the person takes antibiotics
   c. a bacterial cell in the human body that mutates when a person takes antibiotics

8. A fundamental characteristic of life on Earth is that _____.
   a. living things have a complex, organized structure based on inorganic molecules
   b. living things passively acquire materials and energy from their environment and convert them into different forms
   c. living things grow and reproduce
   d. living things reproduce using information stored in RNA
   e. none of the above

9. Identify a characteristic that is representative of some but not all life-forms.
   a. change over time in order to adapt to changing environments
   b. acquisition and use of energy
   c. maintenance of a constant internal environment despite changing external environments
   d. movement over great distances

10. Organisms that can make their own food are called _____; organisms that must obtain energy from molecules made by other organisms are called _____.
   a. herbivores, carnivores
   b. photosynthetic, herbivores
   c. heterotroph, autotroph
   d. autotroph, heterotroph

11. The process of homeostasis is
   a. the mechanism by which living organisms evolve.
   b. how cells produce energy.
   c. how organisms maintain a constant internal environment.
   d. a good name for a rock-and-roll band.

12. The complexity of living systems is
   a. highly organized.
   b. unique to living systems.
   c. the single defining characteristic of living things.
   d. found only in humans.

13. Which of the following BEST describes organisms that contain prokaryotic cells?
   a. Bacteria
   b. Archaea
   c. Plantae
   d. Bacteria and Archaea

14. You discover a new type of organism in the back of your fridge. Luckily, your roommate is a biology major and takes you to the lab where he works. You put a small piece of the fuzzy critter under the microscope and see that it is made of very simple single cells without a nucleus. What type of organism is this MOST likely to be?
   a. Bacterium
   b. Archaean
   c. Protistan
   d. Fungus

15. Studying biology is a worthwhile endeavor because it allows
   a. for a greater appreciation of the natural world.
   b. us to make more informed decisions on how humans should make decisions.
   c. us to make sense of the biological activity that occurs around us every day.
   d. us to make more informed decisions on health and natural resource management issues.
   e. all of the above.

ANSWER KEY

Quiz 1

1.	a		**9.**	a
2.	a		**10.**	c
3.	a		**11.**	d
4.	d		**12.**	d
5.	a		**13.**	c
6.	e		**14.**	d
7.	b		**15.**	a
8.	c			

Quiz 2

1.	b		**9.**	d
2.	c		**10.**	d
3.	e		**11.**	c
4.	a		**12.**	a
5.	d		**13.**	d
6.	a		**14.**	a
7.	b		**15.**	e
8.	c			

CHAPTER 2: ATOMS, MOLECULES, AND LIFE

OUTLINE

Section 2.1 What Are Atoms?

- **Atoms** are the fundamental structural units of matter. Each atom has a central **atomic nucleus**, which contains positively charged **protons** and uncharged **neutrons**. Orbiting the atomic nucleus are negatively charged **electrons** (**Figure 2-1**).
- Atoms are grouped into 92 naturally occurring types (or **elements**) based on the number of protons in the nucleus (i.e., the **atomic number**). Atoms of the same element may have different numbers of neutrons and are referred to as **isotopes**.
- Electrons orbit the nucleus within **electron shells**, each corresponding to a higher energy level. The innermost shell can hold two electrons, while higher energy shells of biologically relevant atoms can hold up to eight (**Figure 2-2**).
- Exciting an electron causes it to jump to a higher energy shell, which releases energy when it falls back to its original shell position (**Figure 2-3**).

Section 2.2 How Do Atoms Interact to Form Molecules?

- **Molecules** are composed of two or more atoms of the same or different elements, while a **compound** is made up of atoms of different elements.
- Stable (or **inert**) atoms have their outermost electron shells filled to capacity. An atom is **reactive** when its outermost electron shell is not full and will attempt to gain stability by gaining, losing, or sharing electrons with other atoms to fill its outer shell (i.e., it will form **chemical bonds**). Chemical bonds are formed (or are broken) as the result of **chemical reactions** that form new substances. The major types of chemical bonds are **ionic bonds**, **covalent bonds**, and **hydrogen bonds** (**Table 2-2**).
- **Ionic bonds** occur when one atom donates an electron to another atom, each filling its outermost electron shell as a result. Doing so causes the donating atom to become positively charged, while the receiving atom becomes negatively charged. These **ions** are then held together by their electrical attraction to each other (**Figures 2-4** and **2-5**).
- **Covalent bonds** occur when atoms share electrons among their outermost shells (**Figure 2-6**). Depending on the atoms involved, covalent bonds can involve equal sharing of electrons (i.e., **nonpolar**, **Figures 2-6** and **2-7a**) or equal sharing of electrons (i.e., **polar**, **Figure 2-7b**). Most biological molecules utilize covalent bonds.
- **Hydrogen bonds** occur when the negatively charged end of one polar molecule is electrically attracted to the positively charged end (containing hydrogen atoms) of another.

Section 2.3 Why Is Water So Important to Life?

- Water acts as an effective **solvent** that can dissolve molecules held together by ionic bonds (**Figure 2-11**) and polar covalent bonds. These molecules are called **hydrophilic**. Molecules with nonpolar covalent bonds typically do not dissolve in water and are called **hydrophobic**.
- Polar water molecules have high **cohesion** due to the hydrogen bonds that form among them, resulting in high **surface tension** at the water's surface. Cohesion allows plants to transport substances through their tissues.
- When water dissociates, hydrogen ions and hydroxide ions are released in equal proportions, forming a **neutral** solution (**Figure 2-14**). Water can react with **acids** to form an excess of hydrogen ions in solution (an **acidic** solution) or can react with **bases** to form an excess of hydroxide ions in solution (a **basic** solution). The degree of acidity is represented on a pH scale (**Figure 2-15**). A **buffer** is a substance that accepts or releases hydrogen ions in response to changes in hydrogen ion concentrations, thus stabilizing pH levels.
- Water moderates the effects of temperature changes on the body because it requires a lot of energy to heat water (a high **specific heat**), evaporate water (a high **heat of vaporization**), and freeze water (a high **heat of fusion**).
- Water forms ice that becomes less dense when it solidifies (**Figure 2-17**). This ensures that ice will form at the surface of the body of water, forming an insulation layer that delays continued freezing of the water beneath it.

LEARNING OBJECTIVES

Section 2.1 What Are Atoms?

- Describe subatomic particles and their relationship to one another.
- Explain the octet rule and how it applies to atomic interactions.

Section 2.2 How Do Atoms Interact to Form Molecules?

- Describe the differences between molecules and compounds.
- Explain what causes an atom to be reactive or inert.
- Discuss the relationship between chemical bonds and chemical reactions.
- Explain how ionic and covalent bonds form.
- Explain how atoms become ions.
- Describe the differences between covalent and ionic bonds.
- Discuss polar and nonpolar covalent bonds and how polarity affects the solubility of each.
- Describe how hydrogen bonds form.
- Discuss free radical formation and their effect on cells.
- Define antioxidants and discuss their role in cells.

Section 2.3 Why Is Water So Important to Life?

- Describe the special properties of water that make life possible.
- Explain the terms *hydrophobic* and *hydrophilic*.
- Discuss how hydrogen bonding results in the cohesion, surface tension, and adhesion properties of water.
- Explain the unique properties of acids and bases.
- Describe how pH is measured.
- Explain the role of buffers in cells.
- Explain what properties of water contribute to its high specific heat.
- Discuss the importance to life of the response of water to changing temperatures.

QUIZ 1

1. The basic structural units of chemistry and life are
   _____.
   a. atoms
   b. electrons
   c. protons
   d. neutrons
   e. molecules

2. Which of the following is NOT a subatomic particle?
   a. proton
   b. neutron
   c. ion
   d. electron

3. The atomic number of carbon is 6. A carbon atom has _____ protons and _____ electrons.
   a. 3, 3
   b. 6, 6
   c. 6, 12
   d. 6, 3

4. All elements are _____
   a. composed of molecules.
   b. a unique form of matter.
   c. found only in living matter.
   d. found only in nonliving matter.

5. How many single covalent bonds can the nitrogen (atomic # = 7) atom form?
   a. 1
   b. 2
   c. 3
   d. 4

6. What happens when an atom ionizes?
   a. It shares one or more electrons with another atom.
   b. It emits energy as it loses extra neutrons.
   c. It gives up or takes up one or more electrons.
   d. It shares a hydrogen atom with another atom.

7. If electrons in water molecules were equally attracted to the hydrogen and oxygen nuclei, water molecules would be
   a. more polar.
   b. less polar.
   c. unchanged.
   d. all of the above.

8. The chemical bonding properties of an atom are determined directly by the
   a. nucleus.
   b. number of protons.

c. number of neutrons.

d. outer shell of electrons.

9. Which of the following is NOT a characteristic of covalent bonds?

a. Result when atoms gain or lose one or more electrons

b. Result when atoms share one or more electrons

c. Are interactions between the outermost electron shells of atoms

d. Are stronger than ionic bonds in water

10. A hypothetical atom has one electron shell that can hold 10 electrons. This atom has seven protons and is electrically neutral. How many covalent bonds can this atom form?

a. 1

b. 2

c. 3

d. 4

11. Which of the following BEST answers the question: Why is water so essential for life?

a. Water is a good solvent.

b. Water moderates the effects of temperature.

c. When water freezes, it forms ice that floats.

d. (a) and (b) are correct.

e. All of the above are correct.

12. Why are hydrophobic molecules, such as fats and oils, unable to dissolve in watery solutions?

a. Water cannot interact with molecules that have polar covalent bonds, as do fats and oils.

b. Water cannot interact with molecules with ionic bonds, such as fats and oils.

c. Water cannot interact with hydrophobic molecules, such as fats and oils, because hydrophobic molecules form hydrogen bonds with each other, excluding the water.

d. Water molecules form hydrogen bonds with each other, excluding the hydrophobic molecules.

13. You are waiting backstage for your cue to come onstage when you notice that you are breathing rapidly and beginning to feel light-headed. As you try to control your anxiety and slow your breathing, you think about what you learned in your biology class this week and realize that your hyperventilation is changing your blood pH. Explain.

Reminder: Blood pH is maintained by carbonate buffer, which is related to the amount of CO_2 you breathe in or out. One way that your body controls the amount of carbonate in your blood is to change the rate of breathing. When you breathe out, you remove CO_2 and lower the amount of carbonic acid in the blood. As a result, the number of H^+ ions in the blood decreases.

Take your time on this one; relating pH to changes in H^+ concentration can be confusing.

a. Rapid breathing decreases my blood's pH, making it more basic.

b. Rapid breathing decreases my blood's pH, making it more acidic.

c. Rapid breathing increases the pH of my blood, making it more basic.

d. Rapid breathing increases the pH of my blood, making it more acidic.

14. Consider a molecule formed entirely by a carbon atom bonded either to itself or to a hydrogen atom. Would this molecule be soluble in water? Why?

a. No, this molecule would be hydrophobic.

b. It depends on the size of the molecule.

c. Yes, the molecule would be hydrophilic.

d. No, this molecule would be hydrophilic.

15. One solution has a pH of 2 and another has a pH of 3. What would be the relative difference in H^+ ion concentration?

a. twofold difference in concentration

b. fivefold difference in concentration

c. tenfold difference in concentration

d. It is impossible to predict the exact difference in concentration.

QUIZ 2

1. Which of the following list of terms is in the correct order of size, from smallest to largest?

a. electron, proton, atomic nucleus, electron shell, atom, molecule

b. proton, electron, atomic nucleus, electron shell, atom, molecule

c. molecule, atom, electron shell, atomic nucleus, proton, electron

d. atomic nucleus, electron, proton, electron shell, atom, molecule

e. electron, proton, atomic nucleus, electron shell, molecule, atom

2. Atoms are the basic building blocks of elements. Each atom, and therefore the elements themselves, is made up of subatomic bits of matter that are electrically negative (electrons), electrically positive (protons), and electrically neutral or uncharged (neutrons). Which subatomic component defines (characterizes) the fundamental nature of each element?

a. electron

b. proton

c. neutron

d. isotope

3. How many electrons would a carbon atom with six protons in its nucleus have?
   a. 6 electrons
   b. 12 electrons
   c. 10 electrons
   d. 18 electrons

4. Isotopes are
   a. atoms with equal numbers of protons and electrons.
   b. atoms with unequal numbers of protons and electrons.
   c. multiple atoms with the same number of protons but different numbers of neutrons.
   d. none of the above.

5. Nonpolar covalent bonds are different from polar covalent bonds because _____.
   a. electrons are shared unequally in nonpolar covalent bonds and are shared equally in polar covalent bonds
   b. electrons are shared equally in nonpolar covalent bonds and are shared unequally in polar covalent bonds
   c. electrons are lost in nonpolar covalent bonds and are gained in polar covalent bonds
   d. electrons are shared in nonpolar covalent bonds and are lost or gained in polar covalent bonds

6. Oxygen atoms have an atomic number of 8. Neon atoms have 10 electrons. Which answer predicts whether or not these atoms are generally reactive (i.e., can form chemical bonds with other atoms)? (Note that this question is not asking whether oxygen can react with neon.)
   a. Both oxygen and neon are not reactive because their outermost electron orbitals are filled.
   b. Both oxygen and neon are reactive because their outermost electron orbitals are not filled.
   c. Oxygen is reactive because its outermost electron shell contains 6 electrons (is not filled); neon is not reactive because its outermost electron shell contains 8 electrons (is filled).
   d. Oxygen is not reactive because its outermost electron shell contains 8 electrons (is filled); neon is reactive because its outermost electron shell contains 2 electrons (is not filled).

7. Imagine that you wanted to make a time capsule in which you would seal important artifacts from your life (pictures, poems, a lock of your baby hair, etc.), to be opened by your heirs 1000 years from now. To prevent these artifacts from decaying, you want to fill the capsule with a gas that would be *least* reactive. Which of these gases would you choose: oxygen gas (O_2), carbon dioxide (CO_2), argon gas (Ar), or hydrogen gas (H_2). (The atomic numbers of the atoms in these molecules are oxygen = 8, carbon = 6, hydrogen = 2, and argon = 18.)
   a. oxygen gas
   b. carbon dioxide
   c. argon gas
   d. hydrogen gas

8. What allows one atom to physically interact with a second atom?
   a. properties of both nuclei
   b. properties of the electrons
   c. electron shells of both atoms
   d. external energy sources

9. Covalent bonds _____.
   a. are strong compared to ionic bonds or hydrogen bonds
   b. share electrons with another atom
   c. may involve the unequal sharing of electrons between two atoms
   d. all of the above

10. An atom other than hydrogen has a single electron in its outermost shell. The MOST LIKELY outcome for this atom in terms of chemical bonding is that the atom will
    a. gain electrons to fill its outermost shell, becoming a negatively charged ion.
    b. lose its outermost electron, becoming an ion with a charge of +1.
    c. share its single electron with another atom.
    d. share enough electrons to fill its outermost shell.

11. Water's ability to act as a "universal" solvent is due to _____.
    a. the fact that there is so much of it in the world around us and in our own bodies
    b. its natural ability to interact with polar molecules such as ions and proteins
    c. the nature of oxygen, which pulls the hydrogen electrons a little closer to it than they are to the two hydrogen atoms
    d. both (b) and (c) above

12. Water greatly resists increases in temperature because it takes great energy to break the
    a. ionic bonds that hold water molecules together.
    b. covalent bonds that hold water molecules together.
    c. huge number of hydrogen bonds that hold water molecules together.
    d. none of the above.

13. A large, complex organic molecule contains a number of polar covalent bonds. Would you expect this molecule to be soluble in water? Why?
    a. No, I would not expect it to be soluble because it does not contain any ionic bonds.
    b. No, I would not expect it to be soluble because it contains polar covalent bonds.
    c. No, I would not expect it to be soluble because it is a large molecule.
    d. Yes, I would expect it to be soluble because it should be hydrophilic.

14. The molecule H_2SO_4 is ionized to H^+ ions and SO_4^- ions in water. Would you predict this to result in a solution that is acidic or basic?
   a. This would be basic, as there would be more OH^- than H^+ ions.
   b. This would be basic, as there would be more H^+ than OH^- ions.
   c. This would be acidic, as there would be more H^+ than OH^- ions.
   d. This would be acidic, as there would be more OH^- than H^+ ions.

15. The high specific heat of water means that living systems are
   a. more dense than nonliving systems.
   b. more resistant to changes in temperature.
   c. made of solid water.
   d. unable to cool themselves by evaporation of water.

ANSWER KEY

Quiz 1

1.	a		**9.**	a
2.	c		**10.**	c
3.	b		**11.**	e
4.	b		**12.**	d
5.	c		**13.**	c
6.	c		**14.**	a
7.	b		**15.**	c
8.	d			

Quiz 2

1.	a		**9.**	d
2.	b		**10.**	b
3.	a		**11.**	d
4.	c		**12.**	c
5.	b		**13.**	d
6.	c		**14.**	c
7.	c		**15.**	b
8.	b			

CHAPTER 3: BIOLOGICAL MOLECULES

OUTLINE

Section 3.1 Why Is Carbon So Important in Biological Molecules?

- **Organic** molecules are composed of both carbon and hydrogen atoms. All other molecules are considered **inorganic**.
- Organic molecules have a tremendous variety of structures and functions, because of the versatility of carbon atoms. A carbon atom may form up to four covalent bonds with other atoms and can thus assume many sizes and complex shapes.
- **Functional groups** of atoms can attach an organic molecule to the carbon backbone and affect its reactivity. Different functional groups impart different chemical properties to organic molecules (**Table 3-1**).

Section 3.2 How Are Organic Molecules Synthesized?

- Organic molecules are composed of small subunits (**monomers**) bonded together to form longer chains (**polymers**).
- Monomers are joined together to form a polymer by **dehydration synthesis**, while polymers are split into individual monomers by **hydrolysis** (**Figures 3-1** and **3-2**).
- Most biological molecules fall into one of four different categories: **carbohydrates**, **lipids**, **proteins**, or **nucleic acids** (**Table 3-2**).

Section 3.3 What Are Carbohydrates?

- **Carbohydrates** are made of carbon, hydrogen, and oxygen in a 1:2:1 ratio. All carbohydrates are composed of one or more **sugar** monomers.
- **Monosaccharides** are carbohydrates composed of a single sugar molecule, with **glucose** being the most common form (**Figure 3-4**). Other monosaccharides include **fructose**, **galactose**, **ribose**, and **deoxyribose** (**Figures 3-5** and **3-6**).
- **Disaccharides** are carbohydrates composed of two sugar monomers joined by dehydration synthesis (**Figure 3-7**). **Sucrose**, **lactose**, and **maltose** are common disaccharides.
- **Polysaccharides** are carbohydrates composed of many sugar monomers. **Starch** is an energy-storage polysaccharide formed by plants (**Figure 3-8**), while **glycogen** is the animal equivalent. **Cellulose** is an important structural carbohydrate in plants (**Figure 3-9**), while the exoskeleton of some invertebrate animals is composed of **chitin** (**Figure 3-10**).

Section 3.4 What Are Lipids?

- **Lipids** contain long regions made almost entirely of carbon and hydrogen that are joined by nonpolar covalent bonds. Lipids are hydrophobic and can act as energy-storage molecules, contribute to cell membrane structure, waterproof plant and animal surfaces, and form hormones.
- Lipids fall into one of three categories: (1) **fats**, **oils**, and **waxes**, (2) **phospholipids**, and (3) **steroids**.
- **Fats**, **oils**, and **waxes** are lipids that have three structural features in common: (1) they contain only carbon, hydrogen, and oxygen; (2) they contain at least one **fatty acid** subunit; and (3) they usually do not form ring structures.
- Fats and oils (also called **triglycerides**), form by the dehydration synthesis of three fatty acid molecules and one **glycerol** molecule (**Figure 3-11**). These lipids contain a large amount of chemical energy and are used as a long-term energy-storage molecule in animals. **Fats** are solid at room temperature because their fatty acids are **saturated** with hydrogen atoms and thus lack double bonds between carbon atoms (**Figure 3-13**). **Oils** are liquid at room temperature because their fatty acids are **unsaturated** with hydrogen atoms, meaning there are double bonds between some carbon atoms (**Figure 3-14**).
- **Waxes** are not a food source but form waterproof coverings over various plant and animal structures.
- **Phospholipids** are lipids composed of a phosphate "head," glycerol backbone, and two fatty acid "tails" (**Figure 3-15**). The phosphate head is polar and is thus ~~hydrophobic~~, while the fatty acid tails are not. Phospholipids form the bulk of the cell plasma membrane.

- **Steroids** are composed of four carbon rings fused together, with various functional groups attached (**Figure 3-16**). Steroids form cholesterol as well some hormones.

Section 3.5 What Are Proteins?

- **Proteins** are composed of one or more **amino acid** monomers (**Figures 3-18** and **3-19**) joined by **peptide bonds** during dehydration synthesis (**Figure 3-20**). Proteins perform many functions due to the diversity of protein structures (**Table 3-3, Figure 3-17**).
- Proteins can have up to four structural levels. The simplest structure, a chain of amino acids, is called the **primary structure**. The amino acid chain can be twisted into a coiled **helix**, resulting in a **secondary structure.** This helix can be folded upon itself in three dimensions, resulting in a **tertiary structure**. Multiple tertiary structure polypeptides can be joined to form one large protein molecule called a **quarternary structure** (**Figure 3-21**).
- The function of a protein is dependent on its three-dimensional shape, so if the shape is altered (i.e., **denatured**), its original function is disrupted.

Section 3.6 What Are Nucleic Acids?

- **Nucleic acids** are composed of chains of **nucleotide** monomers (**Figure 3-24**), in which the phosphate group of one nucleotide is bonded to the five-carbon sugar of the adjacent nucleotide (**Figure 3-25**).
- **Deoxyribonucleic acid** (or **DNA**) is composed of nucleotides containing the sugar deoxyribose. DNA spells out the genetic information used to construct the proteins of each organism (**Figure 3-26**).
- **Ribonucleic acid** (or **RNA**) is composed of nucleotides containing the sugar ribose. RNA carries the code from DNA into the cell cytoplasm and directs protein synthesis.
- Other nucleotides perform a variety of useful functions. **Cyclic nucleotides** play a role in intercellular communication, **adenosine triphosphate** (or **ATP, Figure 3-27**) molecules carry energy from place to place within a cell, and **coenzymes** facilitate cellular metabolism.

LEARNING OBJECTIVES

Section 3.1 Why Is Carbon So Important in Biological Molecules?

- Explain the differences between organic and inorganic molecules.
- Explain the basis for the variety of organic molecules.

Section 3.2 How Are Organic Molecules Synthesized?

- Describe the relationship between monomers and polymers.
- Explain dehydration synthesis and hydrolysis.
- Describe how dehydration synthesis and hydrolysis relate to each other.

Section 3.3 What Are Carbohydrates?

- Discuss the structure of carbohydrates.
- Describe the principal functions of carbohydrates.
- Describe the types of monosaccharides and their functional differences.
- Discuss three important dietary disaccharides.
- Describe two functional uses for polysaccharides.

Section 3.4 What Are Lipids?

- Describe the chemical composition of oils, fats, and waxes.
- Explain the structural and functional differences between saturated and unsaturated fats and oils.
- Describe the structure of phospholipids.
- Describe the structure of a steroid molecule.
- Explain the functional uses of cholesterol.

Section 3.5 What Are Proteins?

- Describe the diverse functions of proteins.
- Describe the structure of an amino acid.
- Explain how amino acid structure provides for protein synthesis.
- Explain the differences between R groups.
- Describe the four levels of protein structure.
- Explain the primary forces that determine the tertiary structure of proteins.
- Describe the causes of protein denaturation and its consequences.

Section 3.6 What Are Nucleic Acids?

- Describe nucleotide structure.
- Explain the structural and functional differences between DNA and RNA.
- Explain the functions of cyclic AMP, ATP, NAD and FAD.

QUIZ 1

1. Characteristics of carbon that contribute to its ability to form an immense diversity of organic molecules include its
   a. tendency to form covalent bonds.
   b. ability to bond with up to four other atoms.
   c. capacity to form single and double bonds.
   d. ability to bond to form extensive, branched, or unbranched carbon skeletons.
   e. all of the above.

2. In chemical reactions, the molecules that take part in the reaction are called *reactants*. Conversely, molecules produced by a chemical reaction are called *products*. During the process of polymerization (synthesis of biological polymers), water is a _____, and the reaction is consequently called a _____ reaction.
   a. reactant, dehydration
   b. reactant, hydrolysis
   c. product, dehydration
   d. product, hydrolysis

3. Carbohydrates _____.
   a. can exist as monomers (monosaccharides), dimers (disaccharides), and polymers (polysaccharides)
   b. have the general chemical formula of $(CH_2O)_n$
   c. can function as a source of energy as an extremely durable structural material, depending upon the specific nature of the chemical bonds between subunits (monomers)
   d. all of the above

4. Complex carbohydrates have numerous roles in plant cells, including providing
   a. short-term energy storage and acting as enzymes.
   b. long-term energy storage and giving structure to plant cell walls.
   c. short-term energy storage and giving structure to plant cell walls.
   d. storing genetic information.

5. The general class of biological molecules that contains large, nonpolar regions that make these molecules insoluble in water is called _____.
   a. phospholipids
   b. fats
   c. lipids
   d. waxes

6. Phospholipids contain a "head group" that is _____ and two fatty acid "tails" that are _____.
   a. hydrophobic, hydrophilic
   b. hydrophilic, hydrophobic
   c. hydrolyzed, nonhydrolyzed
   d. hydrophilic, hydrophilic

7. Proteins are polymers of _____.
   a. peptides
   b. amino acids
   c. nucleotides
   d. sugars

8. Protein functions in cells include _____.
   a. storage and defense
   b. catalysis of biochemical reactions
   c. structure and movement
   d. transport and defense
   e. all of the above

9. A nucleotide is composed of _____.
   a. a sugar and a phosphate group
   b. a phosphate group and a nitrogen-containing base
   c. a sugar and a nitrogen-containing base
   d. a sugar, a phosphate group, and a nitrogen-containing base

10. How are you able to use the energy in the starch of a potato but not in the cellulose of celery?
    a. We eat more potatoes because they taste better than celery.
    b. Our bodies cannot break the bonds in cellulose but can break those in starch.
    c. Cellulose has fructose monomers, and starch has glucose monomers.
    d. Starch normally comes from animals like us, and cellulose comes from plants.

11. A scientist is studying the metabolism of proteins in yeast and wants to follow the formation of proteins from the earliest possible point. In her experiment, she will feed the yeast radioactive nutrients and follow the fate of the radioactivity in the cells. Which of the following atoms will allow her to exclusively follow proteins in the cell?
    a. radioactive carbon
    b. radioactive nitrogen
    c. radioactive oxygen
    d. radioactive sulfur

12. If a lipid is completely saturated, would you predict it would be liquid or solid at room temperature?
    a. All lipids are solid at room temperature.
    b. There is not enough information to answer this question.
    c. I would predict it would be liquid at room temperature.
    d. I would predict it would be solid at room temperature.

13. Lipids are made of fatty acids, which are made of mostly nonpolar bonds. Will these molecules be soluble or insoluble in water?
    a. soluble in water
    b. insoluble in water
    c. hydrophobic
    d. (b) and (c)

14. Rearranging the order of the subunits in which of the following molecules would have the greatest effect on its function?
    a. starch
    b. protein
    c. DNA
    d. (a) and (b)
    e. (b) and (c)

15. The earliest ancient Egyptians buried their dead in pits in the desert. The heat and dryness of the sand dehydrated the bodies relatively quickly, leaving lifelike mummies. The trunk region of the body dehydrates much more slowly than peripheral regions such as the hands or feet. Based on what you know about how biological molecules are broken apart, which regions of a mummy's body would give the most intact pieces of DNA?
    a. liver
    b. fingers
    c. stomach
    d. thigh muscles

QUIZ 2

1. *Organic* is a term we often see in common usage. To a chemist, organic compounds
   a. are compounds of carbon and hydrogen.
   b. are found only in health food stores.
   c. can be made only by living organisms.
   d. are based on a skeleton of oxygen.

2. How are large organic molecules (macromolecules) synthesized?
   a. hydrolysis of monomers
   b. hydrolysis of polymers
   c. dehydration reaction utilizing monomers
   d. dehydration reaction utilizing polymers

3. What is the difference between a simple sugar and a complex carbohydrate?
   a. Complex carbohydrates are polymers of simple sugars.
   b. Sugars are found in proteins and carbohydrates are found in nucleic acids.
   c. Sugars are made by plants and carbohydrates are made by animals.

   d. Carbohydrates are liquid at room temperature and sugars are solid.

4. Foods that are high in fiber are most likely to be derived from
   a. plants.
   b. dairy products.
   c. meat.
   d. fish.
   e. all of the above.

5. Saturated fats _____.
   a. have no double covalent bonds
   b. are solid at room temperature
   c. contain the maximum number of hydrogen atoms possible
   d. (a) and (b) are correct.
   e. (a), (b), and (c) are correct.

6. The quaternary level of protein structure _____.
   a. refers to a functional (biologically active) complex of two or more three-dimensional proteins (e.g., hemoglobin)

    b. is demonstrated by the three-dimensional organization of a single polypeptide

    c. refers to the serial sequence, or arrangement, of amino acids that make up a protein

    d. none of the above

7. Proteins are macromolecules that can perform many different functions within an organism. This diversity of function for this macromolecule is due to the unique nature of one of its four functional groups that is bound to a central carbon atom. Which functional group is responsible for the wide diversity of function attributed to proteins?

    a. carboxyl group

    b. amino group

    c. hydrogen atom group

    d. R group

8. What is the role of DNA and RNA in cells?

    a. preserving and expressing genetic information

    b. forming long fibers such as those in hair

    c. storing energy in fat cells

    d. speeding up chemical reactions as catalysts

9. Which, if any, of the following choices does NOT properly pair an organic compound with one of its building blocks (subunits)?

    a. polysaccharide–monosaccharide

    b. fat–fatty acid

    c. nucleic acid–glycerol

    d. protein–amino acid

    e. all are paired correctly

10. Imagine that you see a diagram of molecular structures in which the atoms are shown as dots and covalent bonds are shown as lines between the dots. The molecules in the diagram have a variety of complex shapes, including long chains, branched chains, rings, and combinations of chains and rings. What type of molecules are these and what is the key atom in them?

    a. They are organic molecules and, therefore, contain carbon.

    b. They are inorganic molecules and, therefore, do not contain carbon.

    c. They are organic molecules and, therefore, do not contain carbon.

    d. They were created in a laboratory and, therefore, contain unnatural atoms.

11. At the gym one day, you notice a new "Energy Bar" being sold that advertises quick energy for your workout. To impress you further, it is claimed that this bar contains only carbon, oxygen, and hydrogen atoms. What kind of biological molecule(s) would you be eating if you ate this "Energy Bar"?

    a. carbohydrates

    b. proteins

    c. DNA and RNA

    d. lipids

    e. carbohydrates and lipids

    f. carbohydrates and proteins

12. You have identified a protein that is unable to form disulfide bridges. Which of the following would this affect?

    a. primary structure of the protein

    b. secondary structure of the protein

    c. tertiary structure of the protein

    d. dehydration synthesis

13. Fat accumulates in our bodies when we overeat because lipids

    a. are important for cell membranes.

    b. have enzymatic activity.

    c. are found only in plants.

    d. are a form of long-term energy storage.

14. Which of the following organic molecules would you expect to be soluble in water?

    a. a carbon backbone with many hydroxyl groups, OH

    b. a carbon backbone with many hydrogens, H

    c. a carbon backbone with methyl groups, CH_3

    d. not enough information to decide, because functional groups do not determine if an organic molecule is water soluble.

15. The liver converts excess nitrogen wastes from our food into urea. Which of the following types of foods would you expect to increase the levels of urea in your blood?

    a. vegetables that are high in carbohydrates

    b. vegetable oils that are high in lipids

    c. meats that contain a lot of protein

    d. legumes that contain a lot of nucleic acid

    e. (a) and (b)

    f. (c) and (d)

ANSWER KEY

Quiz 1

1. e
2. c
3. d
4. b
5. c
6. b
7. b
8. e

9. d
10. b
11. d
12. d
13. b
14. e
15. b

Quiz 2

1. a
2. c
3. a
4. a
5. e
6. a
7. d
8. a

9. c
10. a
11. e
12. c
13. d
14. d
15. f

CHAPTER 4: CELL STRUCTURE AND FUNCTION

OUTLINE

Section 4.1 What Is the Cell Theory?

- **Cell theory** states that (1) all living things are made of one or more cells, (2) the smallest organisms are single cells, and cells are the functional units of multicellular organisms, and (3) all cells arise from preexisting cells.

Section 4.2 What Are the Basic Attributes of Cells?

- Although cells range in size (**Figure 4-1**), the majority are very small as to optimize the movements of nutrients and wastes in or out of them.
- Cells are enclosed by a thin, double layer of phospholipids with embedded proteins called the **plasma membrane** (**Figure 4-2**). It mediates the movements of substances in and out of the cell.
- All cells contain **cytoplasm**, which consists of the material between the plasma membrane and DNA (**Figure 4-3**). The fluid portion of the cytoplasm is called **cytosol**.
- All cells use **deoxyribonucleic acid (DNA)** as genetic material, while **ribonucleic acid (RNA)** is used to copy genes on DNA and to help make proteins.
- All cells obtain energy and nutrients from their environment.

Section 4.3 What Are the Major Features of Eukaryotic Cells?

- **Eukaryotic** cells contain their DNA within a nucleus and possess membrane-bound **organelles** within their cytoplasm (**Figures 4-3** and **4-4**).
- The outer surfaces of plant, fungi, and some protist cells are covered by a supportive and protective **cell wall** (**Figure 4-5**). Plant cell walls are made of cellulose and polysaccharides, fungal cells are made of polysaccharides and chitin, and protist cell walls can be made of cellulose, protein, or silica. Cell walls are typically porous, allowing the passage of chemicals through the cell membrane.
- Organelles are attached to a network of protein fibers called the **cytoskeleton** (**Figure 4-6**). The cytoskeleton helps maintain cell shape, causes cell and organelle movement, and assists in cell division.
- Some cells have bristlelike **cilia** or whiplike **flagella** (**Figure 4-8**) that can propel individual cells or move fluids along their surfaces.
- The **nucleus** of the cell houses the DNA of a eukaryotic cell (**Figure 4-9a**) by enclosing it in a porous, double layered membrane (the **nuclear envelope**). **Chromatin** inside the nucleus forms long **chromosome** strands of DNA (**Figure 4-10**). The **nucleoli** inside the nucleus make **ribosomes** that act as a workbench during protein synthesis (**Figure 4-11**). Ribosomes are distributed within the cytoplasm or along the membranes of the nucleus and endoplasmic reticulum.
- The **endoplasmic reticulum (ER)** is a folded series of membranes that are composed of a smooth and rough portion, named for the presence or absence of ribosomes embedded in the membrane (**Figure 4-12**). The **smooth ER** can manufacture lipids, break down carbohydrates, or metabolize substances depending on the type of cell. The **rough ER** manufactures proteins.
- The **Golgi apparatus** modifies, sorts, and packages the products of the rough ER (as **vesicles**), which then are released from the cell by exocytosis (**Figures 4-13** and **4-14**).
- In some cells, the Golgi apparatus can form **lysosomes**, whose enzymes digest food particles that have been internalized by **food vacuoles** (**Figure 4-15**). Other cells form **contractile vacuoles** for water regulation (such as in freshwater protists, **Figure 4-16**) or **central vacuoles** for support and waste storage (such as in plants, **Figure 5-11**).
- **Mitochondria** are specialized to metabolize food molecules aerobically to make large amounts of ATP as an energy source (**Figure 4-17**). These reactions are associated with a pair of mitochondrial membranes that enclose a fluid matrix.

- **Chloroplasts** are found in photosynthetic eukaryotic cells. They are surrounded by a double membrane inside of which is the green pigment **chlorophyll** (**Figure 4-18**). Chlorophyll captures the energy in sunlight and uses it to make sugars from carbon dioxide and water.
- **Plastids** are used by plants and photosynthetic protists to store sugars that were made during photosynthesis in the form of starches (**Figure 4-19**).

Section 4.4 What Are the Major Features of Prokaryotic Cells?

- Prokaryotic cells are small; have a cell wall; and have rodlike, spherical, or helix shapes. Some move by flagella. Infectious bacteria possess **capsules**, **slime layers**, and/or pili that help them adhere to host tissues (**Figure 4-20a**).
- Prokaryotes lack nuclei and membranous organelles.
- Prokaryotes possess a single, circular **chromosome**, which is coiled in the central region of the cell (the **nucleoid**). Most prokaryotes have small DNA rings (**plasmids**) that carry genes that impart special properties to the cell.
- Prokaryotic cytoplasm contains **ribosomes** that function during protein synthesis.

LEARNING OBJECTIVES

Section 4.1 What Is the Cell Theory?

- Discuss the three principles of cell theory.

Section 4.2 What Are the Basic Attributes of Cells?

- Explain the limitations on cell size.
- Discuss the structure and function of the plasma membrane.
- Describe the components of the cytoplasm.

Section 4.3 What Are the Major Features of Eukaryotic Cells?

- Discuss the structure, function, and location of a eukaryotic cell wall.
- Describe the structural components of the cytoskeleton and their respective functions.
- Explain the structural and functional differences between cilia and flagella.
- Describe the structure and functions for the three major parts of the nucleus.
- Explain the structure and function of the smooth and rough endoplasmic reticulum.
- Describe the structure and functions of the golgi apparatus.
- Describe the structure and functions of lysosomes.
- Discuss how the membranes of organelles compare with the plasma membrane.
- Describe the structure and functions of mitochondria.
- Describe the structure and functions of chloroplasts.
- Describe the structure and functions of the central vacuole.
- Discuss the structural and functional connection between chloroplasts and plastids.

Section 4.4 What Are the Major Features of Prokaryotic Cells?

- Describe the structural components of prokaryotes.
- Describe the differences between prokaryotic and eukaryotic cells.
- Explain the role of plasmids as they relate to pathogenic bacteria.

QUIZ 1

1. All life is made up of
   a. silicon compounds.
   b. cells.
   c. viruses.
   d. pure protein.

2. Which of the following BEST describes the structures found in all cells?
   a. cell wall
   b. plasma membrane
   c. nucleus

d. ribosomes
e. (a) and (c)
f. (b) and (d)

3. Why must living cells remain microscopic in size?
   a. Cells produce a limited number of enzymes.
   b. The energy needs of giant cells would outstrip the available supply of energy from the environment.
   c. Exchanges of substances at the membrane surface would take too long to diffuse throughout the interior of the cell.
   d. (a) and (b) are correct.

4. In what manner do molecules such as proteins and RNA enter into or exit from the nucleus?
   a. diffusion through the lipid bilayers of the nuclear envelope
   b. movement through pores in the nuclear envelope
   c. osmosis through the lipid bilayers of the nuclear envelope
   d. breakdown of the nuclear envelope

5. Chromosomes consist of _____.
   a. DNA
   b. proteins
   c. RNA
   d. proteins and RNA
   e. proteins and DNA

6. Which of the following correctly lists organelles that are part of the internal membrane system of eukaryotic cells?
   a. endoplasmic reticulum, Golgi apparatus, and lysosomes
   b. endoplasmic reticulum and mitochondria
   c. Golgi apparatus and nucleus
   d. endoplasmic reticulum, Golgi apparatus, and cell wall

7. Imagine that you are late for a date and you reach your friend's door out of breath because you just ran the last three blocks from the bus stop. In a lame effort to impress and to try to make your date forget that you are half an hour late, you describe what oxygen is used for in your cells. Which of the following is correct?
   a. The lysosomes in my muscle cells need this extra oxygen to digest sugars and provide me with energy for running.
   b. The cellular enzymes in my leg muscles need this extra oxygen to repair the damage that occurs to my muscle cells as I run.
   c. The mitochondria in my muscle cells need the extra oxygen to produce sugars that, in turn, provide the energy I need to run.
   d. The mitochondria in my muscle cells need this extra oxygen to break down sugars and produce the energy I need to run.

8. Which organelle would you expect to be in abundance in the liver of a drug addict?
   a. Golgi complex
   b. smooth endoplasmic reticulum
   c. rough endoplasmic reticulum
   d. nucleolus
   e. ribosome

9. Which of the following originated by endosymbiosis?
   a. chloroplasts
   b. mitochondria
   c. flagella
   d. lysosomes
   e. (a) and (b) are correct

10. Ribosomes are found in
    a. plant cells.
    b. animal cells.
    c. bacterial cells.
    d. mitochondria.
    e. chloroplasts.
    f. all of the above.

11. The function of the mitochondria is to
    a. contain the genetic material.
    b. synthesize proteins.
    c. convert food into energy for the cell.
    d. package materials for export from the cell.

12. The difference between smooth and rough ER is the presence of
    a. the Golgi complex in RER.
    b. ribosomes on RER.
    c. nuclear pores in RER.
    d. mitochondria on RER.

13. Proteins that are going to be exported from the cell are synthesized _____.
    a. in the Golgi complex
    b. on the rough endoplasmic reticulum
    c. in mitochondria
    d. on the smooth endoplasmic reticulum

14. Prokaryotic cells _____.
    a. are large cells, typically greater than 10 mm in diameter
    b. include numerous membrane-enclosed structures known as organelles
    c. possess a single strand of DNA but no definable membrane-enclosed nucleus
    d. all of the above

15. Plasmids are located _____.
    a. in the nucleus
    b. in the nucleolus
    c. in the cytoplasm
    d. continuous with the nuclear envelope

QUIZ 2

1. The cell theory states that _____.
   a. cells are generally small to allow for diffusion
   b. all cells contain cytoplasm
   c. cells are either prokaryotes or eukarotes
   d. all living things are composed of cells
   e. all cells arise from organic molecules such as DNA

2. The smallest type of cells are mycoplasmas, which have a diameter between 0.1 micrometer and 1.0 micrometer. Mycoplasmas most likely are _____.
   a. fungi
   b. viruses
   c. plant cells
   d. bacteria
   e. animal cells

3. What do we call the fluid in which all cellular organelles are suspended?
   a. cytoplasm
   b. nucleus
   c. endoplasmic reticulum
   d. Golgi complex

4. The endoplasmic reticulum is needed for _____.
   a. synthesis of certain proteins
   b. hormone synthesis
   c. detoxification
   d. synthesis of lipids
   e. all of the above
   f. none of the above

5. Sorting and modification of proteins is an important function of _____.
   a. mitochondria
   b. chloroplasts
   c. lysosomes
   d. the Golgi complex
   e. the plasma membrane

6. Which of the following lists the correct order in which newly synthesized proteins are delivered to the plasma membrane?
   a. endoplasmic reticulum to lysosomes to the Golgi apparatus to the plasma membrane
   b. endoplasmic reticulum to the Golgi apparatus to the plasma membrane
   c. Golgi apparatus to the endoplasmic reticulum to the plasma membrane
   d. endoplasmic reticulum to the plasma membrane
   e. endoplasmic reticulum to the Golgi apparatus to lysosomes to the plasma membrane

7. Which of the following organelle(s) is (are) found in animal cells but not in plant cells?
   a. mitochondria
   b. chloroplasts
   c. central vacuole
   d. cell wall
   e. cytoskeleton
   f. none of the above

8. In certain types of genetic engineering, DNA is injected into the nucleus of a recipient animal cell. What is the fewest number of membranes that must be pierced by the microscopic needle in order to inject the DNA? (Note that the needles used are not small enough to pass through a nuclear pore.)
   a. one
   b. two
   c. three

9. A researcher has discovered an unusual organism deep in the crust of Earth. She wants to know whether it is prokaryotic or eukaryotic. Imagine that she has rapid tests to determine if the following molecules are present: DNA, RNA, and the proteins that form microtubules. You would advise her to test for
   a. DNA, because only eukaryotes have a nucleus.
   b. RNA, because only eukaryotes have ribosomes.
   c. microtubule proteins, because only eukaryotes have microtubules.

10. Lysosomes contain very powerful digestive enzymes that can break down proteins, carbohydrates, and other molecules. Why don't these enzymes digest the cell itself?
    a. The enzymes will digest only foreign material.
    b. The enzymes are separated from the cytoplasm by the lysosomal membrane.
    c. The enzymes are inactive until secreted from the cell.

11. If the nucleus is the control center of the cell, how is information encoded and shipped to the cytoplasm?
    a. by RNA
    b. by chromosomes
    c. by nuclear pores
    d. by the nucleolus

12. Which of the following characteristics of mitochondria are true?
    a. They are able to take energy from food molecules and store it in high-energy bonds of ATP.
    b. All metabolic conversion of high-energy molecules (e.g., glucose) to ATP occurs within the mitochondria.
    c. They are able to convert solar energy directly into high-energy sugar molecules.
    d. They release oxygen during the process of aerobic metabolism.

13. Which of the following generalizations can you make about the cytoskeleton?
    a. The name implies a fixed structure like the bones of a vertebrate or 2×4 boards in the wall of a building.

b. A variety of cytoskeletal elements are integral in the performance of numerous essential cellular functions.

c. It provides a type of cellular armor on the outside of cells that serves a protective function.

d. all of the above

14. Suppose a *Paramecium,* a freshwater protist, were placed in salt water. What would occur (assume the salt water is hypertonic to the *Paramecium*)?

a. The contractile vacuole would need to expel water more frequently.

b. The cell would swell up and burst.

c. The contractile vacuole would shrink.

d. All of the above are correct.

15. Prokaryotic cells do not contain _____.

a. endoplasmic reticulum

b. ribosomes

c. cytoplasm

d. DNA

e. cell walls

ANSWER KEY

Quiz 1

1.	b	**9.**	e
2.	f	**10.**	f
3.	c	**11.**	c
4.	b	**12.**	b
5.	e	**13.**	b
6.	a	**14.**	c
7.	d	**15.**	c
8.	b		

Quiz 2

1.	d	**9.**	c
2.	d	**10.**	b
3.	a	**11.**	a
4.	e	**12.**	a
5.	d	**13.**	b
6.	b	**14.**	c
7.	f	**15.**	a
8.	c		

CHAPTER 5: CELL MEMBRANE STRUCTURE AND FUNCTION

OUTLINE

Section 5.1 How Is the Structure of a Membrane Related to Its Function?

- **Cell membranes** perform five crucial functions: (1) isolation of cell contents from the environment, (2) regulation of the exchange of substances between the cell and extracellular fluid, (3) cell-to-cell communication, (4) attachment within and between cells, and (5) regulation of biochemical reactions.
- Cell membranes are composed of a flexible double layer of phospholipids that contains a shifting patchwork of proteins (the **fluid mosaic model**, **Figure 5-1**).
- Phospholipids are composed of a hydrophilic (polar) phosphate head and two hydrophobic (nonpolar) fatty acid tails (**Figure 5-2**). Phospholipids form a cell membrane **bilayer**, with hydrophobic tails facing each other and hydrophilic heads facing the watery environment outside and inside the cell (**Figure 5-3**).
- Most animal cell membranes also contain cholesterol, which adds strength and flexibility, as well as reducing permeability to water-soluble substances.
- Many types of proteins are found in the cell membrane. **Receptor proteins** bind to specific molecules and trigger reactions within the cell (**Figure 5-5**). **Recognition proteins** are found on cell surfaces and act as identification tags. **Enzymes** catalyze chemical reactions. **Attachment proteins** anchor a cell membrane to another cell or to internal cytoskeleton protein filaments. Transport proteins move hydrophilic molecules through the cell membrane through pores (**channel proteins**) or by binding to the molecule, changing its shape, and depositing it on the other side of the cell membrane (**carrier proteins**).

Section 5.2 How Do Substances Move Across Membranes?

- Molecules in fluids move in response to their **concentration gradient**, from regions of high to regions of low concentration (a process called **diffusion**), until the concentrations are equalized (**Figure 5-6**). Diffusion is more rapid at higher temperatures and greater concentration gradients. Diffusion cannot move molecules rapidly over long distances.
- Transport of molecules across the cell membrane occurs by **passive transport** (does not require energy) or energy-requiring transport.
- Passive transport occurs in a number of ways. Lipid-soluble molecules and dissolved gases move directly through the cell membrane by **simple diffusion** (**Figure 5-7a**). Ions and water-soluble molecules require channel or carrier proteins to diffuse through the cell membrane, a process called **facilitated diffusion** (**Figure 5-7b, c**). Water diffuses across the cell membrane by **osmosis**, which is influenced by the **tonicity** differences across a selectively permeable membrane (**Figures 5-8**, **5-9**, and **5-10**).
- Energy-requiring transport occurs by three mechanisms: (1) **Active transport** uses energy (ATP) and carrier proteins (**pumps**) to move molecules against their concentration gradients (**Figure 5-12**). (2) **Endocytosis** allows a cell to engulf particles of fluids by surrounding them with its cell membrane. Endocytosis of a fluid is called **pinocytosis** (**Figure 5-13**) and endocytosis of a large particle is called **phagocytosis** (**Figure 5-15**). **Receptor-mediated endocytosis** allows a cell to engulf a specific molecule that binds to cell membrane receptor proteins (**Figure 5-14**). (3) Unwanted materials can be expelled from a cell by **exocytosis**, which is the reverse of endocytosis (**Figure 5-16**).
- Generally, most cells are small to optimize their surface area/volume ratios (**Figure 5-17**). This allows them to transport adequate amounts of materials across their cell membranes to sustain themselves.

Section 5.3 How Do Specialized Junctions Allow Cells to Connect and Communicate?

- **Desmosomes** attach adjacent cells together even when stretched (**Figure 5-18a**).
- **Tight junctions** attach cells to each other and prevent substances from leaking between them, forming a waterproof barrier (**Figure 5-18b**).

- **Gap junctions** form channels that join the cytosol between adjacent animal cells (**Figure 5-19a**). These junctions allow cells to communicate easily with each other by allowing specific substances to pass from cell interior to cell interior.
- **Plasmodesmata**, located in the walls of plant cells, are the functional equivalents of gap junctions (**Figure 5-19b**).

LEARNING OBJECTIVES

Section 5.1 How Is the Structure of a Membrane Related to Its Function?

- Describe the major functions of cell membranes.
- Explain the fluid mosaic model of cell membranes.
- Explain how the polar nature of phospholipids gives rise to the membrane structure.
- Describe how the hydrophobic tails of phospholipids contribute to isolating the cell from its environment.
- Describe the functions of the five different types of membrane proteins.

Section 5.2 How Do Substances Move Across Membranes?

- Discuss the relationship among the terms *solute, solvent*, and *concentration*.
- Explain how random motion combined with a concentration gradient results in diffusion.
- Describe the external factors affecting the rate of diffusion.
- Discuss the major difference between passive transport and active transport.
- Explain the basis for the selective permeability of the membrane.
- Describe the differences and similarities among simple diffusion, facilitated diffusion, and osmosis.
- Explain the terms *isotonic, hypertonic*, and *hypotonic* and their relationship to osmosis between a cell and its environment.
- Explain how osmosis and turgor pressure provide support for plants.
- Explain why cells engage in active transport.
- Discuss the various types of endocytosis and exocytosis.
- Relate the process of diffusion to the size limit of a single cell.

Section 5.3 How Do Specialized Junctions Allow Cells to Connect and Communicate?

- Describe the structural and functional differences among desmosomes, tight junctions, and gap junctions.
- Describe the structure and functions of plasmodesmata.

QUIZ 1

1. The fluid mosaic model describes membranes as fluid because the _____.
   a. phospholipids of membranes are constantly moving from one layer to the other layer
   b. membrane is composed mainly of water
   c. phospholipid molecules are bonded to one another, making them more moveable
   d. phospholipids move from place to place around the motionless membrane proteins
   e. phospholipids and proteins move from place to place within the bilayer

2. Recognition proteins function to _____.
   a. regulate the movement of ions across the cell membrane
   b. bind hormones and alter the intracellular physiology of a cell
   c. permit the cells of the immune system to distinguish between pathogens such as bacteria and cells of your own body
   d. all of the above

3. As a result of a high percentage of unsaturated fatty acids in their membranes, certain microorganisms are
   a. able to withstand high temperatures because their membranes are firmer.
   b. able to withstand low temperatures because their membranes do not solidify as rapidly.
   c. able to carry out sodium ion transport more efficiently than most organisms.
   d. more rapidly recognized and destroyed by cells of the immune system.

4. Which of the following types of molecules must pass through membranes via the aqueous pores formed by membrane proteins?
    a. gases such as carbon dioxide and oxygen
    b. water
    c. large particles such as bacteria
    d. small charged ions such as Na^+ and Ca^{++}

5. Diffusion is the movement of molecules from _____.
    a. an area of higher concentration of that type of molecule to an area of lower concentration
    b. an area of lower concentration of that type of molecule to an area of higher concentration
    c. outside the cell to inside the cell

6. In osmosis, water diffuses from the side of the membrane with a higher concentration of water to the side with a lower concentration of water. What determines the concentration of water in a solution?
    a. the volume of the solution
    b. the amount of molecules other than water dissolved in the solution
    c. the size of the container

7. Which of the following processes does a cell use to take up molecules against their concentration gradient?
    a. simple diffusion
    b. facilitated diffusion
    c. active transport
    d. endocytosis
    e. (c) and (d)

8. When a drop of food coloring is placed in a glass of water, the spreading out of the molecules of food dye is caused by _____.
    a. the concentration of the molecules of dye
    b. the movement of molecules to a lower concentration
    c. the random movement of molecules
    d. osmosis

9. When placed in a certain sucrose solution, the volume of a cell decreases; therefore, the sucrose solution is _____ to the cell contents.
    a. isotonic
    b. hypertonic
    c. hypotonic
    d. cannot determine from given information

10. Which of the following substances will diffuse most rapidly across the plasma membrane?
    a. amino acid
    b. sodium ion
    c. water
    d. oxygen

11. What will happen if there is a high concentration of oxygen on one side of a biological membrane?
    a. The oxygen will stay on the one side because it cannot cross the membrane.
    b. The oxygen will diffuse across the membrane until the concentration of each side is the same.
    c. The oxygen will be actively transported across the membrane.

12. A chloride ion is in a cell that has no channel proteins in its membrane. Will the chloride be able to get out of the cell?
    a. Yes, the chloride ion has a concentration gradient from inside to outside, so it can diffuse out of the cell.
    b. No, there is no channel to diffuse through, and a charged ion or molecule cannot diffuse down its concentration gradient without a channel to pass it through the hydrophobic portion of the membrane.
    c. No, the chloride ion can only follow a concentration gradient to get into the cell.
    d. No, the chloride already entered the cell because of a concentration gradient.

13. How will the water react when a cell is placed in a hypotonic solution?
    a. Water will move out of the cell.
    b. Water will move into the cell.
    c. There will be no net movement of water.
    d. It is impossible to predict the movement of water.

14. Two types of connections between cells called *gap junctions* and *plasmodesmata* are specialized to _____.
    a. prevent the movement of molecules between cells that are tightly joined along ribbons of cell membrane
    b. tightly hold one cell against another at focal points, almost like a spot weld of superglue
    c. permit the passage of substances (e.g., ions) between cells through small passageways that directly link the cytoplasm of one cell to the cytoplasm of another cell
    d. none of the above

15. Which cell junction functions to make the connections between cells leakproof?
    a. desmosomes
    b. tight junctions
    c. gap junctions
    d. plasmodesmata

QUIZ 2

1. Membrane fluidity within a phospholipid bilayer is based upon _____.
   a. interactions among nonpolar (hydrophobic) lipid tails
   b. hydrophilic interactions among polar phospholipid heads
   c. the presence of transport proteins in the lipid bilayer
   d. the presence of water in the lipid bilayer

2. A hormone circulating in the bloodstream would most likely bind to _____.
   a. a recognition protein
   b. a receptor protein
   c. a channel protein
   d. protein filaments in the cytoplasm

3. The plasma membrane of the cell is
   a. permeable to water.
   b. impermeable to water.
   c. permeable to all charged molecules.
   d. permeable to all large molecules.

4. What is the difference between active transport and passive transport?
   a. Passive transport involves the movement of substances directly through the lipid portion of a membrane. Active transport requires an input of energy, whereas passive transport does not.
   b. Active transport requires energy and is unable to move substances against their concentration gradient. Passive transport does not require energy and can move substances against their concentration gradient.
   c. Active transport requires energy and can move substances against their concentration gradient. Passive transport does not require energy and can move substances only down their concentration gradient.

5. Following is a list of substances. Each substance needs a different type of transport to get INTO a cell. Pick the choice below that accurately matches the type of transport needed to convey each substance INTO a cell.

   Substance list: oxygen, water, sodium ions, potassium ions, Bacterium. (Note that sodium ions are more concentrated outside cells than inside; potassium ions are more concentrated inside cells than outside.)
   a. simple diffusion, osmosis, facilitated diffusion, active transport, exocytosis
   b. osmosis, simple diffusion, facilitated diffusion, active transport, endocytosis
   c. passive transport, osmosis, simple diffusion, facilitated diffusion, active transport

   d. simple diffusion, osmosis, facilitated diffusion, active transport, endocytosis
   e. endocytosis, active transport, active transport, facilitated diffusion, exocytosis

6. Imagine that you are studying cell structure in various organisms in your biology lab. Your instructor gives you a microscope slide showing two types of cells that have been suspended in pure water. One type of cell swells up until it bursts. The other cell maintains its shape throughout the experiment. Suggest an explanation for these observations. Assume that both cells were alive at the start of the experiment and that the concentration of water inside both types of cells is similar.
   a. The cell that burst lacked gap junctions, so water that entered the cell via osmosis could not leak back out through the junctions.
   b. The cell that burst lacked a plasma membrane for regulating osmosis.
   c. The cell that remained intact had plasmodesmata that allowed the excess water to leak out, thus balancing the tendency of water to enter the cell via osmosis.
   d. The cell that remained intact had a contractile vacuole for pumping out the excess water that entered the cell via osmosis.

7. Facilitated diffusion requires _____.
   a. a membrane transport protein
   b. a concentration gradient
   c. energy
   d. (a) and (b)

8. During endocytosis, the contents of the endocytic vesicle _____.
   a. enter the cell
   b. exit the cell
   c. enter or exit the cell, always moving down a concentration gradient

9. Which of the following transport processes require(s) energy?
   a. facilitated diffusion
   b. osmosis
   c. endocytosis
   d. facilitated diffusion and osmosis
   e. facilitated diffusion, osmosis, and endocytosis

10. The point at which a substance is evenly dispersed within a fluid _____.
   a. is referred to as a *dynamic equilibrium*
   b. occurs when random movements of the substance cease
   c. happens when net diffusion is zero
   d. (a) and (c)

11. Substances are able to cross the lipid bilayer of a cell at different rates that are unique for each substance. Which of the following characteristics would favor the simple diffusion of a substance across a cell membrane?
    a. low lipid solubility
    b. small molecule size
    c. small concentration gradients
    d. the number of membrane transport proteins

12. Many metabolic poisons work by inhibiting ATP production. Which type of transport would be most affected?
    a. osmosis
    b. facilitated diffusion
    c. active transport
    d. simple diffusion

13. The concentration of sodium ions is lower in the cytoplasm of a heart muscle cell than it is in the extracellular fluid. By what mechanism does the cell maintain this difference?
    a. active transport
    b. osmosis
    c. facilitated diffusion
    d. endocytosis

14. Which of the following characteristics describes cell walls?
    a. porous
    b. strong
    c. composed of cellulose or polysaccharides
    d. all of the above
    e. none of the above

15. Which of the following associations is NOT correct?
    a. Gap junctions—allow communication between animal cells
    b. Plasmodesmata—allow communication between plant cells
    c. Desmosomes—allow inflexible attachments among cells
    d. Tight junctions—create watertight junctions between cells.

ANSWER KEY

Quiz 1

1. e
2. c
3. b
4. d
5. a
6. b
7. e
8. c
9. b
10. d
11. b
12. b
13. b
14. c
15. b

Quiz 2

1. a
2. b
3. a
4. c
5. d
6. d
7. d
8. a
9. c
10. d
11. b
12. c
13. a
14. d
15. c

CHAPTER 6: ENERGY FLOW IN THE LIFE OF A CELL

OUTLINE

Section 6.1 What Is Energy?

- **Energy** is defined as the capacity to do **work**, which is the force acting on an object that causes it to move. **Kinetic energy** is the energy of movement, while **potential energy** is stored energy. Energy can be changed from potential to kinetic energy (and vice versa) under the right conditions (**Figure 6-1**).
- The basic properties of energy are described by the **laws of thermodynamics**. The **first law of thermodynamics** (i.e., the **law of conservation of energy**) states that energy can neither be created nor destroyed, but it can change forms. The second law of thermodynamics states that when energy changes forms, the amount of useful energy decreases (**Figure 6-2**).

Section 6.2 How Does Energy Flow in Chemical Reactions?

- **Chemical reactions** form or break chemical bonds between atoms by converting **reactants** into **products**. All chemical reactions either release energy (are **exergonic**, **Figure 6-3**) or require energy (are **endergonic**, **Figure 6-4**). All chemical reactions require **activation energy** to begin, usually in the form of kinetic energy from moving molecules (**Figure 6-6**).
- **Coupled reactions** occur when an exergonic reaction provides the energy needed for an endergonic reaction to occur. In cells, this often occurs through the use of **energy-carrier molecules**.

Section 6.3 How Is Cellular Energy Carried Between Coupled Reactions?

- **Adenosine triphosphate** (**ATP**) is the most common energy-carrier molecule used by cells (**Figure 6-8**). ATP releases energy from its high-energy phosphate bonds when it is broken down into **adenosine diphosphate** (**ADP**) and phosphate (**Figure 6-10**). Energy is stored when phosphate forms chemical bonds with ADP (**Figure 6-9**). Heat is given off during energy transfers between exergonic and endergonic reactions (**Figure 6-11**).
- **Electron carriers** can also transport energy by transferring energetic electrons within cells (**Figure 6-12**).

Section 6.4 How Do Cells Control Their Metabolic Reactions?

- The metabolic reactions of a cell are linked by **metabolic pathways** (**Figure 6-13**) that synthesize or break down molecules.
- Spontaneous reactions that sustain life proceed too slowly at body temperatures due to the high activation energies required. **Enzymes** are proteins that act as biological **catalysts** that reduce the activation energy (and increase the speed), of these reactions (**Figure 6-14**).
- An enzyme functions when one or more **substrate** molecules fit into its **active site**. The substrates and active site change shape, promoting a reaction that forms a new molecular product (**Figure 6-15**).
- Cells regulate metabolic reactions by controlling enzymes synthesis, by synthesizing enzymes in inactive forms, or by producing regulatory molecules that change enzyme activity (called **allosteric regulation**, **Figure 6-16**). **Feedback inhibition** is a form of allosteric regulation in which a metabolic pathway stops producing a product when it reaches a specific level (**Figure 6-17**).
- Some poisons and drugs compete with substrates for the active site of an enzyme, disrupting that specific metabolic pathway (called **competitive inhibition**, **Figure 6-18**).
- Enzymes function optimally within narrow temperature and pH ranges (**Figure 6-19**) because suboptimal conditions disrupt their three-dimensional protein structure.

LEARNING OBJECTIVES

Section 6.1 What Is Energy?

- Describe the types of kinetic and potential energies.
- Discuss the law of conservation of energy as it relates to kinetic and potential energy.
- Explain how the second law of thermodynamics does not violate the law of conservation of energy.
- Describe the sun's role in the second law of thermodynamics.

Section 6.2 How Does Energy Flow in Chemical Reactions?

- Explain the relationship of reactants to products in chemical reactions.
- Discuss the role of energy in exergonic and endergonic reactions.
- Explain the role of activation energy in chemical reactions.
- Describe photosynthesis in terms of the conservation of mass and energy.
- Explain how exergonic and endergonic reactions in a cell are coupled via ATP.

Section 6.3 How Is Cellular Energy Carried Between Coupled Reactions?

- Describe the structure of ATP and how it is able to serve as an energy-carrier molecule.
- Discuss how using ATP results in an increase in temperature.
- Describe the role of NAD^+ and FAD in energy transfer within a cell.

Section 6.4 How Do Cells Control Their Metabolic Reactions?

- Describe how enzymes increase the rate of a reaction.
- Describe the role of enzymes in a metabolic pathway.
- Explain the role of coenzymes.
- Explain what factors determine the specificity of an enzyme.
- Explain how a cell regulates its metabolism through enzymes.
- Discuss allosteric regulation and its role in feedback inhibition.
- Describe how competitive inhibition affects the functioning of enzymes.
- Describe the effects of temperature, salt concentration, and pH on enzymatic function.

QUIZ 1

1. The laws of thermodynamics define the properties and behavior of energy. The first law states that energy _____.
   a. equals mass times the speed of light, squared (i.e., $E = mc^2$)
   b. can be created by thermonuclear explosions
   c. cannot be created or destroyed but can be changed from one form into another
   d. is the basic structure of the universe

2. Which of the following does NOT illustrate kinetic energy?
   a. a car moving at 55 mph
   b. a 100-watt light bulb, turned on
   c. a 9-volt battery
   d. a water droplet going down a waterfall
   e. the electrical current used to perk coffee

3. When electrical energy is used to turn on a light bulb, the conversion from electrical energy to light energy is not 100% efficient. This loss of usable energy can be explained by _____.
   a. the first law of thermodynamics

   b. the second law of thermodynamics
   c. a destruction of energy
   d. a conversion to potential energy

4. The second law of thermodynamics relates the organization of matter to energy. It states that unless additional energy is used, the orderliness of a system tends to _____, whereas entropy _____.
   a. increase, decreases
   b. decrease, increases
   c. stay the same, increases
   d. decrease, stays the same

5. In a chemical reaction, the _____ are the atoms or molecules that enter into the reaction; the _____ are the chemicals or atoms produced by the reaction.
   a. products, reactants
   b. reactants, products
   c. receptors, products
   d. catalysts, reactants

6. In endergonic reactions, _____.
   a. the reactants have more potential energy than the products

b. energy is released
c. a net input of energy is not required
d. all of the above
e. none of the above

7. Activation energy is _____.
   a. required for endergonic reactions
   b. produced by exergonic reactions
   c. required for all chemical reactions
   d. produced by chemical reactions

8. Which of the following situations illustrates the coupling of exergonic to endergonic reactions in cells?
   a. the production of ATP by breakdown of glucose
   b. the active transport of sodium into the cell
   c. the movement of a muscle powered by the hydrolysis of ATP
   d. all of the above

9. ATP is well suited to its role as an energy-carrier molecule in cells because _____.
   a. the covalent bond between the last two phosphates can be broken to release substantial amounts of energy
   b. it contains covalent bonds
   c. it is small and can fit into a lot of places in the cell
   d. the covalent bonds between the last two phosphates are high-energy bonds that can absorb a substantial amount of energy when the bonds are broken

10. Electron carrier molecules that transport energy include
    a. NAD^+.
    b. FAD.
    c. ADP.
    d. CO_2.
    e. (a) and (b).

11. The speed of a reaction is determined by its _____.
    a. reactants
    b. products
    c. activation energy
    d. potential energy

12. Biological catalysts _____.
    a. are organic substances that lower the activation energy required to initiate and speed up a reaction
    b. are broken down and destroyed as part of the chemical process they help initiate
    c. are seldom regulated by the molecules that participate in the catalyzed reactions
    d. can speed up any reaction, even those that would not occur naturally

13. Pepsin is an enzyme that breaks down protein but will not act upon starch. This fact is an indication that enzymes are _____.
    a. catalytic
    b. hydrolytic
    c. specific
    d. temperature sensitive

14. Enzyme regulation can be precisely controlled through a variety of mechanisms such as _____.
    a. producing an inactive form of an enzyme that is activated only when needed
    b. using a coenzyme that is necessary for function
    c. binding a competitive inhibitor to the active site
    d. all of the above

15. Lysosomes have an acidic interior (pH = 5), unlike the rest of the cell (pH = 7). Lysosomal enzymes are most active at _____.
    a. pH = 4
    b. pH = 5
    c. pH = 7
    d. pH = 9

QUIZ 2

1. If all matter tends toward increasing randomness and disorder, how can life exist?
   a. Living things do not obey the second law of thermodynamics.
   b. There is a constant input of energy from the sun.
   c. Living things do not require energy.
   d. All of the above are correct.

2. The laws of thermodynamics _____.
   a. explain how energy is created and destroyed
   b. refer to fluctuating temperature changes
   c. define the basic properties and behavior of energy
   d. suggest that matter in a disorganized state requires energy to be constantly added to maintain the "disorganization"

3. The second law of thermodynamics states that
   a. energy transactions are completely efficient.
   b. energy cannot be created or destroyed.
   c. disorder tends to increase over time.
   d. energy is constantly being created.

4. Potential energy represents
   a. the energy of motion.
   b. stored energy.
   c. newly created energy.
   d. energy that is unavailable for use.

5. If the products of a reaction have more energy than the reactants, then we say this is an
   a. exergonic reaction.
   b. endergonic reaction.

6. Which kind of reaction will occur spontaneously?
   a. exergonic reactions
   b. endergonic reactions

7. The importance of coupled reactions is that they
   a. allow an endergonic reaction to drive an exergonic reaction.
   b. allow an exergonic reaction to drive an endergonic reaction.
   c. turn potential energy into kinetic energy.
   d. violate the second law of thermodynamics.

8. A coupled reaction is one in which
   a. a couple of reactants are brought together to form a product.
   b. one reaction immediately follows another reaction.
   c. energy from an exergonic reaction is used to drive an endergonic reaction.
   d. a couple of reactions occur close to one another in the cell.

9. ATP is important to cells because it
   a. is exergonic and can be coupled to endergonic reactions.
   b. is endergonic and can be coupled to exergonic reactions.
   c. lowers the activation energy for an exergonic reaction.

10. How is ATP used in energy metabolism?
    a. ATP synthesis is coupled to the endergonic reactions of photosynthesis.
    b. ATP synthesis results from the reactions of protein synthesis.
    c. ATP synthesis is coupled to an exergonic reaction, the breakdown of glucose.
    d. ATP hydrolysis increases entropy.

11. Which of the following are correct statements about how enzymes work?
    a. Enzymes catalyze specific chemical reactions because the shape of their active site allows only certain substrate molecules to enter.
    b. The substrate(s) enter the enzyme active site in specific orientations.
    c. The enzyme active site and the substrate(s) change shape, promoting a specific chemical reaction and then allowing the products to leave.
    d. (a), (b), and (c)
    e. (a) and (b) are correct

12. The most important reason a particular enzyme can function only within certain limits of temperature, salt conditions, and pH is that changes in temperature, salt, and pH _____.
    a. change the shape of an enzyme
    b. alter the amount of substrates
    c. alter the amount of products
    d. lower the activation energy of a reaction

13. A substance that is acted upon by an enzyme to produce a product is called a(n) _____.
    a. allosteric inhibitor
    b. coenzyme
    c. substrate
    d. electron carrier

14. The amino acid threonine is converted to isoleucine by a sequence of five enzymatic reactions. When isoleucine levels are high, the first reaction in this sequence is "turned off." This is an example of _____.
    a. substrate activation
    b. feedback inhibition
    c. competitive inhibition
    d. coenzyme activation

15. Why do most reactions occur more rapidly at high temperature?
    a. Molecules move more rapidly at higher temperatures.
    b. Collisions between molecules are more frequent.
    c. Collisions will be hard enough to force electron shells to interact.
    d. All of the above are correct.

ANSWER KEY

Quiz 1

1. c
2. c
3. b
4. b
5. b
6. e
7. c
8. d

9. a
10. e
11. c
12. a
13. c
14. d
15. b

Quiz 2

1. b
2. c
3. c
4. b
5. b
6. a
7. b
8. c
9. a
10. c
11. d
12. a
13. c
14. b
15. d

CHAPTER 7: CAPTURING SOLAR ENERGY:
PHOTOSYNTHESIS

OUTLINE

Section 7.1 What Is Photosynthesis?

- In **photosynthesis**, autotrophic cells capture light energy (in the presence of CO_2 and H_2O) and store it in the chemical bonds of glucose, releasing O_2 in the process. The energy in glucose is released for use by the cell by the process of **cellular respiration** (**Figure 7-1**).
- **Leaves** are flattened structures that expose a large surface area to the sun. This allows **chloroplasts** to collect sunlight for use in photosynthesis (**Figure 7-2**).
- Two chemical reactions occur during photosynthesis: (1) **light-dependent reactions** that convert sunlight energy and convert it to energy-carrier molecules and (2) **light-independent reactions** that convert the energy in carrier molecules into glucose (**Figure 7-4**).

Section 7.2 Light-Dependent Reactions: How Is Light Energy Converted to Chemical Energy?

- Sunlight photons are captured by **chlorophyll** and accessory pigment molecules that allow chloroplasts to absorb light at different wavelengths (**Figure 7-5**).
- Light-dependent reactions occur within the **thylakoid membranes** of chloroplasts in specialized **photosystems** (**Figure 7-7**). **Photosystem II** generates ATP **chemiosmotically** by passing an energized electron along an **electron transport chain**. This electron becomes reenergized by **photosystem I**, which passes it to an electron transport chain that forms NADPH. (**Figure 7-8**).
- Some of the energy formed from these reactions is used to split H_2O, which provides O_2, H^+, and oxygen for use in photosynthesis and cell respiration.

Section 7.3 Light-Independent Reactions: How Is Chemical Energy Stored in Glucose Molecules?

- ATP and NADPH formed during the light-dependent reactions dissolve in the **stroma** fluid and are used by light-independent reactions to synthesize glucose from CO_2 and H_2O.
- This process begins with the **Calvin-Benson cycle** (C_3 **cycle**), which has three major parts: (1) carbon fixation, (2) synthesis of G3P, and (3) regeneration of RuBP (**Figure 7-10**). During **carbon fixation**, plants capture CO_2 and fix its atoms in a larger molecule (RuBP), which then splits into two PGA molecules. During **synthesis of G3P**, energy in ATP and NADPH is used to convert PGA into G3P. During **regeneration of RuBP**, ATP is used to assemble RuBP from G3P. ADP and $NADP^+$ are formed during this process.
- Two G3P molecules formed during the C_3 cycle are combined to form one glucose molecule. Glucose is then used to form sucrose, starch, or cellulose.

Section 7.4 What Is the Relationship Between Light-Dependent and Light-Independent Reactions?

- The light-dependent reactions use light to form energy-carrier molecules (ATP and NADPH), which are then used by the light-independent reactions (C_3 cycle) to drive glucose synthesis.
- The ADP and $NADP^+$ formed during the C_3 cycle are used by the light-dependent reactions to reform ATP and NADPH.

Section 7.5 Water, CO_2, and the C_4 Pathway

- To conserve water, leaf stomata close, causing leaf CO_2 levels to drop and leaf O_2 levels to rise. This causes the enzyme **rubisco** to combine O_2 with RuBP (**photorespiration**), which prevents carbon fixation and does not generate ATP (**Figure 7-12a**).
- C_4 plants minimize photorespiration by adding an additional step to the carbon-fixation process. In mesophyll cells, CO_2 is combined with **phosphoenolpyruvic acid (PEP)** to form a four-carbon molecule, which releases CO_2 after being transported to adjacent bundle-sheath cells. This CO_2 is then fixed using the C_3 cycle.
- C_4 pathways use more energy than C_3 pathways. Thus, C_3 plants have an advantage in cool, wet, low-light environments, whereas C_4 plants have an advantage in hot, dry, well-lit environments.

LEARNING OBJECTIVES

Section 7.1 What Is Photosynthesis?

- Describe the reactants and products of photosynthesis.
- Discuss the structural features of leaves involved in photosynthesis.
- Describe the structure and function of the chloroplast.
- Describe the general differences between the light-dependent and light-independent reactions of photosynthesis.

Section 7.2 Light-Dependent Reactions: How Is Light Energy Converted to Chemical Energy?

- Explain how chlorophyll a, chlorophyll b, and the carotenoids work together to absorb the light necessary for photosynthesis.
- Explain the difference among absorption, reflection, and transmission of light as it relates to photosynthesis.
- Explain why leaves change colors in the fall.
- Explain how the structure of the thylakoid provides for its light-energy-absorbing function.
- Describe how photosystem II uses light and electrons to generate ATP.
- Describe the flow of electrons from photosystem II to photosystem I.
- Explain the role of $NADP^+$ in photosystem I.

Section 7.3 Light-Independent Reactions: How Is Chemical Energy Stored in Glucose Molecules?

- Describe the three stages of the Calvin-Benson cycle.
- Explain the ATP connection between the light-dependent and light-independent reactions.

Section 7.4 What Is the Relationship Between Light-Dependent and Light-Independent Reactions?

- Discuss the interrelationship between the light-dependent and light-independent reactions of photosynthesis.

Section 7.5 Water, CO_2, and the C_4 Pathway

- Describe the conditions that lead to photorespiration and its consequences for the plant.
- Explain the C_4 pathway with respect to the plant's environment.

QUIZ 1

1. In most land plants, photosynthesis occurs in cells of the _____ of the leaves, because these cells contain the largest numbers of chloroplasts.
   a. epidermis
   b. stomata
   c. cuticle
   d. mesophyll
   e. vascular bundles

2. Which of the following is a TRUE statement about photosynthesis?
   a. In photosynthesis, inorganic molecules such as carbon dioxide and water react to produce organic, energy-rich molecules such as glucose.
   b. In photosynthesis, oxygen is used to help break down glucose.
   c. Photosynthesis is an exergonic reaction.
   d. Photosynthesis is a process that is carried out primarily by autotrophic prokaryotic bacteria.

3. When photosystem I absorbs a photon, electrons are ejected from the reaction center. What is the final destination of these electrons?
   a. The electrons are used to replace ones lost from photosystem II.
   b. Electrons end up being captured by the electron carrier NADPH.
   c. They are accepted directly by ATP.
   d. They were last seen at the tables in Vegas.

4. How are the electrons lost from photosystem II replaced?
   a. They come from photosystem I.
   b. They come directly from energetic photons.
   c. They diffuse through the thylakoid membrane.
   d. They result when water splits with oxygen as by-product.

5. Light reactions take place in the _____.
   a. thylakoid membrane
   b. stroma
   c. space within the thylakoid disk
   d. mitochondria

6. You have just discovered a new plant with red-orange leaves. Which wavelengths of visible light are NOT being absorbed by this pigment?
   a. green, blue, and violet
   b. green and yellow
   c. green
   d. red and orange
   e. blue and violet

7. In the light-dependent reactions of photosynthesis, the difference in hydrogen ion concentration across the thylakoid membrane is used to generate _____.
   a. NADPH
   b. glucose
   c. $FADH_2$
   d. ATP
   e. oxygen

8. Light energy is initially captured by *photosystems* within thylakoid membranes. Photosystems are organized arrays of _____.
   a. proteins
   b. chlorophyll molecules
   c. pigment molecules such as carotenoids
   d. all of the above
   e. none of the above

9. Which of the following is NOT required for the light-independent reactions of photosynthesis?
   a. ATP
   b. stroma
   c. NADPH
   d. CO_2
   e. H_2O
   f. O_2

10. The dark reactions of photosynthesis
    a. convert carbon dioxide to sugar.
    b. generate ATP and NADPH.
    c. make oxygen.
    d. use the atoms in ATP to make sugar.

11. Dark reactions take place in the
    a. thylakoid membrane
    b. space within thylakoid disks
    c. mitochondria
    d. stroma

12. Which of the following represents the products of the light-dependent reactions of photosynthesis? For which process are these products needed?
    a. ATP and oxygen; photosystem I
    b. NADPH and glucose; light-independent reactions
    c. carbon dioxide and water; Calvin-Benson cycle
    d. NADH and ADP; cellular respiration
    e. NADPH and ATP; Calvin-Benson cycle

13. Why is the use of carbon dioxide in the dark reactions important?
    a. The carbon dioxide produced by the light reactions is used by the dark reactions.
    b. The plant would not be able to get rid of carbon dioxide if the dark reactions did not use it.
    c. Fixation of carbon dioxide in the dark reactions makes it usable for all life forms.
    d. The dark reactions convert carbon dioxide directly into ATP to make energy.

14. Photorespiration is the process by which
    a. plants produce energy at night.
    b. plant cells cool off in hot climates.
    c. sugar production is prevented in C_3 plants when CO_2 levels are low and O_2 levels are high.
    d. plants capture light energy and convert it into ATP.

15. How are C_4 plants different from C_3 plants?
  a. C_4 plants function better in low-light conditions.
  b. C_4 plants function better in relatively moist conditions.
  c. C_4 plants function better in bright-light conditions.
  d. C_4 plants function better in relatively dry conditions.
  e. (c) and (d)

QUIZ 2

1. Leaves include a number of structural modifications for the purpose of photosynthesis, including _____.
  a. adjustable openings in the surface that permit the passage of CO_2, H_2O, and O_2
  b. a waxy covering on the surface of the leaf that is designed to reduce evaporation
  c. photosynthetic mitochondria
  d. (a) and (b)

2. Light-independent, carbon-fixing reactions occur in
  a. guard cell cytoplasm
  b. chloroplast stroma
  c. the thylakoid membranes
  d. the thylakoids
  e. mitochondria

3. During photosynthesis, electrons are continuously lost from the reaction center of photosystem II. What source is used to replace these electrons?
  a. sunlight
  b. oxygen
  c. water
  d. carbon dioxide

4. In the light-dependent reactions of photosynthesis, ATP is produced by chemiosmosis. Describe this process.
  a. Chemiosmosis is the process by which water moves across a semipermeable membrane.
  b. As high-energy electrons move from carrier to carrier in the electron transport system of the thylakoid membrane, some of the energy is captured to pump hydrogen ions across it.
  c. When light strikes the chlorophyll molecules, water is moved via osmosis across the chloroplast membrane. When water moves back across the membrane, ATP is generated.

5. A lovely tree called the *flowering plum* has beautiful pink flowers in spring and deep-purple leaves in summer. What types of photosynthetic pigments are plentiful in the leaves of this plant? Can photosynthesis occur in these purple leaves? Explain your answer.
  a. The leaves probably contain a high amount of phycocyanins and carotenoids and less chlorophyll than do plants with green leaves. They can still perform photosynthesis, because photosynthesis can occur to some extent at all wavelengths of light.
  b. The leaves probably contain a high amount of chlorophyll and carotenoids and lesser amounts of phycocyanins than do plants with green leaves.
  c. The leaves probably contain a high amount of carotenoids and lesser amounts of phycocyanins and chlorophyll than do plants with green leaves.

6. Take a deep breath and slowly exhale. The oxygen in that breath is being used by your mitochondria in reactions that produce ATP from sugars and other food molecules you ate. Where did that oxygen come from originally?
  a. The oxygen in the atmosphere is produced by the breakdown of carbon dioxide during photosynthesis.
  b. Plants produce the oxygen via the process of photorespiration when they break down sugars in their mitochondria.
  c. The oxygen in the air is produced by the splitting of water during the light-dependent reactions of photosynthesis.
  d. Plants produce the oxygen via the process of photorespiration, in which they break down sugars at night when photosynthesis cannot occur.

7. Imagine that you are trying to set up a large fish tank and want to select a colored light that will show off the fish to best advantage but will also allow the green plants in the tank to grow and stay healthy. You decide to measure the efficiency of photosynthesis by looking at the production of oxygen bubbles on the leaves (see Figure 7-5). At first, you use a white fluorescent lamp and see many oxygen bubbles on the leaves, indicating that photosynthesis is occurring normally. Next, you put a sheet of green cellophane between the white light and the water so that the light coming through to the tank appears green. Now, the bubbles will
  a. remain constant because photosynthesis can occur to some extent at all visible wavelengths of light.
  b. increase because chlorophyll is green, and photosynthesis will work best if leaves are exposed to green light.
  c. decrease because green light cannot be absorbed by chlorophyll.
  d. disappear because photosynthesis cannot occur in green light.

8. The role of electron transport in ATP synthesis is to
   a. Electron transport creates a proton gradient that is used to supply energy for the synthesis of ATP.
   b. Electron transport chains donate the final electrons directly to ATP.
   c. Electron transport chains donate electrons to phosphate, which transfers them to ATP.
   d. Electron transport chains are necessary to split water to generate oxygen.

9. Why is carbon dioxide a key molecule in the light-independent reactions of photosynthesis?
   a. Carbon dioxide provides electrons to replace those lost by chlorophyll during the light-dependent reactions.
   b. Carbon dioxide, with water, is the raw material for the synthesis of sugars, the key products of these reactions.
   c. Carbon dioxide inhibits the light-independent reactions of photosynthesis.
   d. Carbon dioxide is the major product of the light-independent reactions of photosynthesis.

10. A scientist studying photosynthesis illuminated a culture of algae with bright visible light. She then turned out the light and simultaneously began to bubble radioactive CO_2 gas into the culture. After 30 minutes, she stopped the reaction and measured the amount of radioactivity inside the cells. What did she find? Explain your answer.
    a. There was no radioactivity inside the cells because the CO_2 is used to produce O_2 in the light-dependent reactions. Thus, there was radioactivity in the air above the culture but not in the cells.
    b. There was radioactivity in the cells because the CO_2 is used to synthesize sugar, even in the dark.
    c. There was no radioactivity in the cells because light is required to produce sugars from CO_2 and water.
    d. There was radioactivity inside the cells, because CO_2 is used to replace the electrons that were lost by chlorophyll when the lights were turned on.

11. The term *carbon fixation* refers to _____.
    a. the loss of carbon during glucose synthesis
    b. the synthesis of glyceraldehyde-3-phosphate
    c. the incorporation of atmospheric carbon dioxide into a larger organic molecule
    d. the regeneration of ribulose bisphosphate

12. What is the linkage between the light and dark reactions?
    a. The light reactions are endergonic and the dark reactions are exergonic
    b. The light reactions produce ATP and NADPH, and the dark reactions require ATP and NADPH.
    c. The light reactions pass electrons via an electron transport chain to the dark reactions.
    d. The light reactions produce carbon dioxide and the dark reactions use carbon dioxide.

13. The _____ reactions "charge-up" ADP and $NADP^+$ energy carriers, while the _____ reactions use the "charged" energy carriers to make glucose.
    a. light-dependent, light-independent
    b. light-independent, light dependent
    c. light-dependent, dark
    d. light-independent, dark

14. Photorespiration occurs when _____.
    a. oxygen is combined with ribulose bisphosphate
    b. carbon dioxide is combined with ribulose bisphosphate
    c. stomata are closed
    d. (a) and (c)

15. How have C_4 plants adapted to environmental conditions that would result in increased photorespiration?
    a. C_4 plants have substituted phosphoenolpyruvate (PEP) for ribulose bisphosphate.
    b. PEP specifically combines with carbon dioxide even in the face of high oxygen concentrations.
    c. A molecule transports the fixed carbon into a cell type where the normal C_3 synthetic metabolism would be favored due to the now high concentrations of carbon dioxide.
    d. all of the above

ANSWER KEY

Quiz 1

1. d
2. a
3. b
4. d
5. a
6. d
7. d
8. d

9. f
10. a
11. d
12. e
13. c
14. c
15. e

Quiz 2

1. d
2. b
3. c
4. b
5. a
6. c
7. c
8. a
9. b
10. b
11. c
12. b
13. a
14. d
15. d

CHAPTER 8: HARVESTING ENERGY: GLYCOLYSIS AND CELLULAR RESPIRATION

OUTLINE

Section 8.1 How Do Cells Obtain Energy?

- Cells commonly obtain energy by breaking the chemical bonds of glucose molecules and using this energy to form **ATP**. This process is called **glucose metabolism** (**Figure 8-1**).
- The first step of glucose metabolism (**glycolysis**) occurs in the cytosol and breaks down glucose into pyruvate and two ATP. If oxygen is absent, **fermentation** converts pyruvate to either lactate or ethanol and CO_2.
- If oxygen is present (aerobic conditions), eukaryotic cell mitochondria undergo **cellular respiration** to break pyruvate into CO_2 and H_2O, forming significantly more ATP than glycolysis (34-36 ATP).

Section 8.2 How Is the Energy in Glucose Captured During Glycolysis?

- Glycolysis occurs in two steps: (1) glucose activation and (2) energy harvest (**Figure 8-2**).
- During **glucose activation**, two ATP react with glucose to form **fructose bisphosphate**, which is an unstable "activated" molecule.
- During **energy harvest**, fructose bisphosphate splits into two **glyceraldehyde-3-phosphate** (**G3P**) molecules that are converted into two **pyruvate** molecules, which results in the production of four ATP and two NADH carrier molecules.
- Under anaerobic conditions, **fermentation** reactions convert pyruvate into either lactate or ethanol and CO_2 (**Figure 8-3**). Fermentation is necessary to produce NAD^+ electron-carrier molecules, which are required for glycolysis to occur (**Figures 8-4** and **8-5**).

Section 8.3 How Does Cellular Respiration Capture Additional Energy from Glucose?

- Pyruvate is first broken down in the mitochondrial matrix in two stages: (1) acetyl CoA formation and (2) the Krebs cycle (**Figure 8-7**).
- During **acetyl CoA formation**, each pyruvate reacts with coenzyme A, splitting it into CO_2 and forming acetyl CoA. NADH is also formed during this process.
- During the **Krebs cycle**, each acetyl CoA combines with oxaloacetate to form citrate. Citrate is then processed by mitochondrial enzymes, regenerating oxaloacetate and producing two CO_2, one ATP, and four electron-carrier molecules (three NADH and one $FADH_2$).
- The electron-carrier molecules produced in glycolysis and mitochondrial matrix reactions (NADH and $FADH_2$) deposit their electrons in the **electron transport chains** (**ETC**) that are embedded in the inner mitochondrial matrix. The deposited electrons move along the chain, generating energy used to pump H^+ to the intermembrane space of the mitochondria, forming a H^+ gradient across the inner mitochondrial membrane. Electrons that reach the end of the ETC are accepted by hydrogen and oxygen to form water, clearing space on the ETC for the movement of additional electrons. The resulting H^+ gradient across the inner mitochondrial membrane drives the **chemiosmotic** production of 32 or 34 ATP by **ATP-synthesizing enzymes** (**Figure 8-8**).

Section 8.4 Putting It All Together

- The mechanisms and energy harvest of glucose metabolism are summarized in **Figures 8-9** and **8-10**.

LEARNING OBJECTIVES

Section 8.1 How Do Cells Obtain Energy?

- Describe the relationship between the products and reactants of photosynthesis and glucose metabolism.
- Explain glycolysis in terms of products and reactants.

Section 8.2 How Is the Energy in Glucose Captured During Glycolysis?

- Describe the two stages of glycolysis in terms of ATP use and production.
- Explain the role of NADH production in glycolysis.
- Explain the reasons for fermentation and its results for the organism.
- Describe lactic acid fermentation in terms of which cells utilize it and why they do so.
- Discuss alcoholic fermentation in terms of which organisms utilize it and describe its products.

Section 8.3 How Does Cellular Respiration Capture Additional Energy from Glucose?

- Describe the structure and functions of the mitochondria.
- Explain the Krebs cycle in terms of reactants, products, and by-products.
- Describe the location and function of the electron transport chain.
- Explain how H^+ ions are utilized to produce ATP.

Section 8.4 Putting It All Together

- Describe the flow of energy transfer from glucose to ATP in aerobic respiration.

QUIZ 1

1. In eukaryotic cells, glycolysis occurs in the _____, and cellular respiration occurs in the _____.
   a. mitochondria, cytoplasm
   b. cytoplasm, mitochondria
   c. cytoplasm, chloroplasts
   d. chloroplasts, mitochondria

2. Photosynthesis and glucose metabolism are related because the _____.
   a. products of photosynthesis are the raw materials for glucose metabolism
   b. products of glucose metabolism are the raw materials for photosynthesis
   c. products of photosynthesis are the same as the products of glucose metabolism
   d. raw materials of photosynthesis are the same as the raw materials of glucose metabolism
   e. (a) and (b)

3. You are comparing two cultures of cells, one that is undergoing cellular respiration and one that is fermenting. Both cultures are producing ATP at the same rate. If this is true, what else would you observe about the fermenting culture?
   a. It would require more glucose per minute than the respiring culture.

   b. It would have a higher rate of glycolysis than the respiring culture.
   c. It would produce pyruvate at a faster rate than the respiring culture.
   d. It would require more oxygen than the respiring culture.
   e. (a), (b), and (c) are correct.

4. Which molecules are produced in glycolysis and used in fermentation?
   a. acetyl CoA and NADH
   b. pyruvate and NADH
   c. glucose, ATP, and NAD^+
   d. lactate, ATP, and CO_2
   e. pyruvate and ATP

5. How many molecules of ATP would be produced from 20 molecules of glucose at the end of fermentation?
   a. 10
   b. 20
   c. 30
   d. 40
   e. 100

6. Which product of the fermentation of sugar by yeast in bread dough is essential for the rising of the dough?
   a. lactate
   b. ATP
   c. ethanol
   d. CO_2
   e. O_2

7. What is the role of fermentation in glucose metabolism?
   a. Fermentation extracts the maximum ATP from glucose.
   b. Fermentation is necessary under aerobic conditions.
   c. Fermentation regenerates NAD^+ under anaerobic conditions.
   d. Fermentation allows the cell to use glucose under aerobic conditions.

8. Which role does the electron transport chain play during cellular respiration?
   a. The electron transport system takes energy from the high-energy electrons brought by electron carriers (e.g., NADH) and uses it to pump hydrogen ions against their concentration gradient from the matrix into the intermembrane compartment.
   b. The electron transport system allows hydrogen ions to diffuse down their concentration gradient from the intermembrane compartment to the matrix.
   c. The electron transport system produces ATP.
   d. All of the above are correct.

9. Suppose that the reactions of mitochondria of a green plant were completely inhibited. Which process would immediately stop?
   a. glycolysis
   b. fermentation
   c. photosynthesis
   d. ATP production
   e. lactate production

10. In eukaryotic cells, pyruvate produced by glycolysis is transported into the mitochondrial matrix, where _____.
    a. enzymes for the Krebs cycle break down the pyruvate, producing CO_2 as a waste product
    b. the electron transport system recombines pyruvate molecules to produce glucose
    c. enzymes for the Krebs cycle convert the pyruvate into alcohol or lactate
    d. the electron transport system breaks down the pyruvate, producing CO_2 as a waste product

11. In eukaryotes, during the process of chemiosmosis, ATP is produced as hydrogen ions move from _____ to _____, passing through _____.
    a. the intermembrane compartment, the matrix, an ATP synthase
    b. the matrix, the intermembrane compartment, an ATP synthase
    c. the cytoplasm, the matrix, the electron transport system
    d. the matrix, the cytoplasm, the Krebs cycle

12. Which products of the Krebs cycle feed the electron transport chain?
    a. NADH and $FADH_2$ (flavin adenine dinucleotide)
    b. pyruvate and acetyl CoA
    c. fructose bisphosphate
    d. carbon dioxide

13. How is electron transport related in chloroplasts and mitochondria?
    a. Electron transport is used in each case to synthesize ATP, although the electron donors and final electron acceptors differ in the two systems.
    b. They have no similarities at all.
    c. The chloroplast system depends on intact membranes while the mitochondrial does not.
    d. Only the mitochondrial electron transport system can synthesize ATP.

14. What is the final electron acceptor in the electron transport chain?
    a. NADH
    b. carbon dioxide
    c. acetyl CoA
    d. oxygen

15. What is the importance of ATP production?
    a. Even though the production of ATP occurs within individual cells, a multicellular organism requires the energy produced to carry out vital functions essential for survival. Any organism would quickly die without constant production of ATP.
    b. Understanding glycolysis and cellular respiration permits insights into how different organisms manage their energy needs in different environments.
    c. (a) and (b) are correct.

QUIZ 2

1. Which of the following statements concerning fermentation is (are) true?
   a. Fermentation requires oxygen.
   b. Fermentation, like glycolysis, occurs in the cytoplasm of cells.
   c. Fermentation produces additional ATP.
   d. The end product of fermentation in human cells is ethanol.
   e. All of the above are correct.
   f. None of the above is correct.

2. The overall equation for glucose metabolism is $C_6H_{12}O_6 + 6\,O_2 \rightarrow 6\,CO_2 + 6\,H_2O + ATP$ and heat. The carbon atoms in the CO_2 molecules in this equation come from _____ during reactions of _____.
   a. O_2, glycolysis
   b. O_2, the electron transport system
   c. O_2, the Krebs cycle
   d. $C_6H_{12}O_6$, glycolysis
   e. $C_6H_{12}O_6$, the electron transport system
   f. $C_6H_{12}O_6$, the Krebs cycle

3. At the beginning of most recipes for bread, you are instructed to dissolve the yeast in a mixture of sugar (sucrose) and hot water, in some cases with a small amount of flour. Within a short time, this yeast mixture begins to bubble and foam, perhaps to the point of overflowing the container. What is happening?
   a. The bubbles are carbon dioxide that yeast produce as they break down the glucose and produce ATP via fermentation.
   b. The bubbles are oxygen produced by yeast as they grow.
   c. The bubbles are detergents that yeast produce to help them digest the proteins in the flour.
   d. The bubbles are water vapor produced as the hot water evaporates.

4. The energy-harvesting reactions of glycolysis produce four molecules of _____, two molecules of _____, and two molecules of _____.
   a. ATP, glyceraldehyde-3-phosphate, pyruvate
   b. ATP, NADH, pyruvate
   c. glucose, carbon dioxide, water
   d. pyruvate, glyceraldehyde-3-phosphate, water

5. What are the two parts of glycolysis?
   a. glucose activation and energy harvest
   b. fermentation and cellular respiration
   c. extracellular and intracellular events

6. Cells recycle NADH back to NAD^+ during fermentation by converting pyruvate to
   a. lactic acid.
   b. ethanol and carbon dioxide.
   c. acetyl CoA.
   d. (a) and (b).

7. The production of which molecule marks the end of glycolysis and the beginning of cellular respiration?
   a. Coenzyme A (CoA)
   b. acetyl CoA
   c. citrate
   d. pyruvate

8. In aerobic organisms growing in the presence of oxygen, the NADH produced by glycolysis ultimately donates its high-energy electrons to _____.
   a. electron transport chains in the mitochondria
   b. glucose
   c. pyruvate
   d. ATP

9. Respiration is the process of gas exchange (breathing in oxygen and breathing out carbon dioxide); cellular respiration is the process of _____.
   a. cellular gas exchange
   b. cellular cooling
   c. production of ATP via the electron transport system
   d. cellular reproduction

10. Oxygen that is breathed in during respiration is used in cellular respiration by being
    a. converted into CO_2 and exhaled.
    b. converted into ATP.
    c. the final electron acceptor of the electron transport system.
    d. used to produce glucose.

11. In eukaryotic cells, the enzymes for the Krebs cycle are located in the _____, and those for the electron transport system are located in the _____.
    a. cytoplasm, cell wall
    b. cytoplasm, mitochondrial matrix
    c. plasma membrane, cytoplasm
    d. mitochondrial matrix, inner mitochondrial membrane
    e. inner mitochondrial membrane, matrix

12. As high-energy electrons are passed from carrier to carrier along the electron transport system in cellular respiration, the electrons lose energy. Some of that energy is *directly* used to _____.
    a. synthesize glucose
    b. break down glucose
    c. pump hydrogen ions across a membrane
    d. synthesize ATP

13. Why does death result from any situation that prevents a person from breathing?
    a. Oxygen is needed for cellular respiration, so lack of oxygen prevents cells from making sufficient ATP for essential cellular functions. Cells die as a result, eventually leading to death of the individual.

b. Glycolysis requires oxygen in order to produce ATP, so lack of oxygen prevents cells from making sufficient ATP.

c. Oxygen is necessary for both fermentation reactions and cellular respiration. so lack of oxygen prevents cells from making sufficient ATP.

14. When glucose goes through glycolysis and all the way through the Krebs cycle, the six carbon molecules of glucose
   a. become pyruvate and G3P.
   b. become carbon dioxide.

c. become joined to NADH.
d. are used to make more glucose.

15. Which pathway produces the most ATP per glucose molecule?
   a. fermentation
   b. glycolysis
   c. Krebs cycle
   d. cellular respiration
   e. electron transport and chemiosmosis

ANSWER KEY

Quiz 1

1. b		**9.** d	
2. e		**10.** a	
3. e		**11.** a	
4. b		**12.** a	
5. d		**13.** a	
6. d		**14.** d	
7. c		**15.** c	
8. a			

Quiz 2

1. b		**9.** c	
2. f		**10.** c	
3. a		**11.** d	
4. b		**12.** c	
5. a		**13.** a	
6. d		**14.** b	
7. d		**15.** d	
8. a			

CHAPTER 9: DNA: THE MOLECULE OF HEREDITY

OUTLINE

Section 9.1 How Did Scientists Discover That Genes Are Made of DNA?

- By the early 1900s, it was established that genetic information existed in units called **genes**, which are parts of **chromosomes**. Since chromosomes were composed of both protein and DNA, it was unclear which acted as the cell's hereditary blueprint.
- Griffith's experiments on mice determined that genes could be transferred from one strain of bacteria to another, causing it to become deadly to mice (**Figure 9-1**).
- Avery, MacLeod, and McCarty determined that this transformation was caused by DNA, thus they concluded that genes are made of DNA (**Figure 9-2**).

Section 9.2 What Is the Structure of DNA?

- DNA is composed of **nucleotides**, each with a phosphate group, a sugar (deoxyribose), and one of four possible nitrogen-containing **bases**: **adenine (A)**, **guanine (G)**, **thymine (T)**, and **cytosine (C)** (**Figure 9-3**).
- Nucleotides are arranged in strands, with the phosphate group of one nucleotide bonded to an adjacent sugar group of another, forming a **sugar-phosphate backbone**. Nitrogen-containing bases protrude from this sugar-phosphate backbone.
- Two DNA nucleotide strands are held together by hydrogen bonds between nitrogen-containing bases of adjacent nucleotide strands, forming a **double-helix** (**Figure 9-5**). Hydrogen bonds form between nitrogen-containing bases in a predictable way: adenine with thymine (A-T) and guanine with cytosine (G-C). These pairings are called **complementary base pairs**.

Section 9.3 How Does DNA Encode Information?

- DNA encodes information based on the sequence of nucleotide bases in a strand, rather than its length.

Section 9.4 How Does DNA Replication Ensure Genetic Constancy During Cell Division?

- During cell division, DNA synthesizes an exact copy of itself through a process called **DNA synthesis**.
- During DNA synthesis, **DNA helicases** pull apart the parental DNA double helix, exposing its nitrogen-containing bases. **DNA polymerases** read the exposed nitrogen-containing bases and bond complementary free nucleotides to them, forming two new DNA strands (**Figure 9-6**).
- In each new DNA double helix, one strand came from the parent DNA double helix while the other is a daughter strand (**semiconservative replication, Figure 9-7**).

Section 9.5 How Do Mutations Occur?

- **Mutations** are changes in the sequence of bases in DNA. Mutations are rare, occurring once every 100 million to 1 billion base pairs. This is due to DNA repair enzymes that proofread each daughter strand during and after its synthesis.
- Mutations can occur from normal errors in base pairing, as well as by environmental conditions that can cause damage to DNA (such as UV rays in sunlight).
- **Nucleotide substitutions** (**point mutations**) occur when the correct nucleotide in a pair is replaced instead of the incorrect one (**Figure 9-8a**). **Insertion mutations** occur when one or more nucleotide pairs are inserted into the DNA double helix (**Figure 9-8b**), while **deletion mutations** occur when pairs are removed from the DNA double helix (**Figure 9-8c**).

- **Inversions** are mutations that occur when chromosome pieces are cut out, turned around, and reinserted (**Figure 9-8d**). **Translocations** are mutations that occur when a chunk of DNA is removed from a chromosome and moved to another one (**Figure 9-8e**).
- Most mutations are harmful or neutral, but some can be beneficial and favored by natural selection, depending on the environment.

LEARNING OBJECTIVES

Section 9.1 How Did Scientists Discover That Genes Are Made of DNA?

- Describe bacterial transformation using Griffith's experiments.
- Explain how Avery's experiments proved DNA and not protein caused transformation.

Section 9.2 What Is the Structure of DNA?

- Explain Chargaff's rules and their implication for the orientation of bases in DNA.
- Describe the information Wilkins and Franklin discovered from X-ray diffraction of DNA.
- Describe the Watson and Crick model of DNA structure.
- Explain how Chargaff's rules fit with the Watson and Crick model of DNA.

Section 9.3 How Does DNA Encode Information?

- Explain how four bases are able to encode all the variety of life.

Section 9.4 How Does DNA Replication Ensure Genetic Constancy During Cell Division?

- Explain the role of complementary base pairing in DNA replication.
- Describe the individual roles of DNA helicases and DNA polymerases in replication.

Section 9.5 How Do Mutations Occur?

- Explain how mutations arise and the mechanisms cells use to minimize their occurrence.
- Describe the outside forces that damage DNA.
- Explain the different types of mutations.
- Discuss the effect of mutation on an organism.

QUIZ 1

1. Identify the research that first provided the basis for the following statement: DNA is the genetic material.
   a. Watson and Crick proposed a new model for DNA structure.
   b. Avery, MacLeod, and McCarty isolated the material that transformed R-strain bacteria into S-strain bacteria.
   c. Chargaff found that DNA contains equal amounts of adenine and thymine, as well as equal amounts of cytosine and guanine.
   d. Wilkins and Franklin used X-ray diffraction to study DNA structure.

2. An interpretation of Griffith's experiments is that
   a. fragments of DNA containing genes were taken up by the R-strain bacteria.
   b. genetic material must have been transferred from the R-strain bacteria into the S-strain bacteria.

   c. the genetic material must be protein.
   d. base-pairing accounts for the amounts of each base found.

3. DNA structure can be described as a twisted ladder. Imagine you are climbing a model of DNA, just as if you were climbing a ladder. Which parts of a nucleotide are your *feet* touching as you climb?
   a. sugars
   b. nitrogenous bases
   c. phosphates

4. Human chromosomes range in size dramatically, with the smallest (sex chromosome Y) being many times smaller than the largest (autosomal chromosome 1). What is responsible for determining the size of a chromosome?
   a. the length of the DNA molecule in it

b. the amount of protein associated with it

c. the number of DNA molecules in it

5. Imagine that you are studying a newly discovered bacterium from a hot springs in Yellowstone National Park. When you examine the nucleotide composition of this organism, you find that 10% of the nucleotides in its DNA are adenine. What percentage of nucleotides are guanine? Explain.

a. 10%, because A pairs with G.

b. 90%, because 100% minus 10% equals 90%, because A pairs with T.

c. 40%, because A pairs with T (accounting for 20% of the bases), leaving 80% of the nucleotides as G-C base pairs; half of 80% is 40%.

6. Which of the following statements is TRUE of the sugar-phosphate backbone of DNA?

a. The two sugar-phosphate strands are oriented in the same direction.

b. The two sugar-phosphate strands are oriented in the opposite direction.

c. The two sugar-phosphate strands can be oriented in either the same or the opposite direction.

d. The sugar-phosphate backbone is in the center of the DNA molecule.

7. In a DNA helix, all of the following are true EXCEPT

a. the nitrogenous bases are covalently bonded to one another.

b. the nitrogenous bases are in the inner part of the helix.

c. the strands are in opposite orientation.

d. cytosine pairs with guanine.

8. Information in DNA is carried in _____.

a. the sugar-phosphate backbone of one DNA strand

b. the base pairs between nucleotides in the two DNA strands

c. the proteins that bind to the DNA double helix

d. the order of the nucleotide bases in one DNA strand

9. Multiple replication bubbles on a single eukaryotic chromosome

a. allow for rapid replication of eukaryotic DNA.

b. do not occur.

c. lead to many DNA strands being synthesized from the same chromosome simultaneously.

d. occur to ensure that the DNA is replicated faithfully.

10. DNA polymerase can

a. replicate both strands of DNA in a continuous manner.

b. add nucleotides only to the free sugar end of a DNA molecule.

c. bond short stretches of DNA together.

11. DNA ligase has all of the functions listed below EXCEPT:

a. repair of damaged DNA.

b. piecing together of small pieces of DNA during replication.

c. covalently bonding DNA strands made by adjoining replication forks.

d. covalently bonding small DNA pieces together during the synthesis of both strands of a double helix.

12. Which of the following statements is FALSE?

a. DNA replication involves uncoiling of the parental DNA molecule.

b. DNA replication produces a long, continuous strand and a series of short pieces.

c. DNA ligase is required in the synthesis of one strand.

d. Both parental strands end up in the same daughter strand after replication.

e. DNA polymerase molecules move toward the replication fork on both strands.

13. The purpose of DNA replication is to produce

a. two similar DNA double helices differing in a small number of specific sites.

b. two very different DNA double helices.

c. two identical DNA double helices.

d. one copy that is identical to the parental DNA molecule and one that is totally different.

e. a single-stranded DNA molecule from the double-stranded parent DNA.

14. Which of the following can cause errors to accumulate in DNA?

a. high levels of metabolic activity

b. ultraviolet light

c. cold temperatures

d. none of the above

15. The approximate error rate for DNA polymerase is one error for every

a. 100 bases added.

b. 10,000 bases added.

c. billion bases added.

45

QUIZ 2

1. Before it was actually determined, many scientists had trouble believing that DNA was the genetic material. This is most likely because
   a. it was known that proteins could be passed from generation to generation.
   b. the number of nucleotides in DNA is very small.
   c. destruction of proteins prevented genetic transformation.
   d. it was known that RNA could be passed from generation to generation.

2. How can a cell killed by heat in order to render it harmless somehow still act to transform a second strain of bacterium from a noninfectious form into a disease carrier (i.e., a pathogen)?
   a. The second (noninfectious) strain of bacterium was changed into a disease carrier (i.e., pathogen) by something from the heat-killed cell.
   b. The heat-killed cell was not really dead.
   c. The second (noninfectious) strain spontaneously mutated into a deadly pathogen.
   d. (b) and (c) are correct.

3. The structure of DNA explained Chargaff's observations because
   a. the DNA molecule is a regular, repeating molecule.
   b. the twisting of the DNA required certain numbers of bases.
   c. adenines were found to pair with thymines and cytosines with guanines.
   d. the sequence of bases is crucial to the storage of information.
   e. there are only four different bases.

4. The two strands of DNA that make up a double helix are
   a. identical to each other.
   b. held together by covalent bonds.
   c. oriented in the same direction.
   d. complementary to each other.

5. Consider the backbone of the DNA double helix. Which statement is NOT true?
   a. It is composed of alternating sugar and phosphate groups.
   b. The backbone is not straight, but twisted.
   c. The DNA backbone forms the central core of the DNA molecule.
   d. There is directionality to the backbone.

6. The sugars and phosphates in the backbone of a DNA strand are held together by _____.
   a. covalent bonds
   b. hydrogen bonds
   c. ionic bonds

7. The two strands of a DNA double helix are held together by _____.
   a. covalent bonds between the sugars of one nucleotide and the phosphates of the adjacent nucleotide
   b. hydrogen bonds between bases on opposite DNA strands
   c. ionic bonds between DNA and water

8. Which of the following lists the correct order of events in DNA replication?
   a. Enzymes unwind the DNA double helix; DNA polymerase makes two new DNA strands complementary to the old ones; the two DNA molecules wind up into a double helix with one new strand and one old strand.
   b. Two DNA molecules wind up into a double helix with one new strand and one old strand; DNA polymerase makes two new DNA strands complementary to the old ones; enzymes unwind the DNA double helix.
   c. DNA polymerase makes the new DNA strands complementary to the old ones; enzymes unwind the DNA double helix; the two DNA molecules wind up into a double helix with one new strand and one old strand.

9. Which of the following events occurs within a DNA replication bubble?
   a. DNA polymerase helps break hydrogen bonds between complementary base pairs.
   b. DNA helicase attaches the phosphate of a free nucleotide to the sugar of the previous nucleotide in the daughter strand.
   c. DNA helicase unwinds the double helix at each replication fork within a replication bubble.
   d. none of the above

10. Which of the following options would result from the actions of DNA polymerase during DNA replication?
    a. Two DNA polymerase molecules act to synthesize a long continuous daughter DNA strand from each parental strand; ligase is not needed.
    b. Two DNA polymerase molecules act to synthesize a short segment of daughter DNA from each parental strand; ligase is used to connect these short segments of both daughter strands.
    c. Two DNA polymerase molecules act to synthesize daughter DNA strands: one via a long continuous strand that moves in the same direction as the helicase, and a second polymerase synthesizes short segments of DNA that must be joined by ligase.
    d. None of the above is correct.

11. Replication bubbles

    a. consist of one moving replication fork and one fixed replication fork.

    b. are always shrinking in size.

    c. are only present once per chromosome.

    d. consist of two replication forks moving in opposite directions.

12. The correct nucleotides to be added to the new DNA strand are determined by

    a. DNA ligase.

    b. base pairing between the free nucleotides and bases on the template strand.

    c. DNA helicase.

    d. base pairing between the free nucleotides and bases on the new DNA strand.

13. DNA replication is a semiconservative process because

    a. each of the resulting DNA double helices will consist of one newly synthesized strand and one parental strand.

    b. the resulting DNA molecules will consist of one with two newly synthesized strands, and the other will contain the original parental strands.

    c. the nucleotides from other nucleic acids are constantly being recycled to make new nucleic acids.

    d. each resulting DNA molecule will contain short stretches of newly synthesized DNA interspersed with the original parental DNA.

14. The approximate error rate for DNA replication is one error for every

    a. thousand nucleotides.

    b. million nucleotides.

    c. billion nucleotides.

15. Which of the following can be caused by ultraviolet radiation in sunlight?

    a. Adjacent thymines become linked together.

    b. The DNA molecule is degraded from each end.

    c. DNA molecule becomes fragmented.

    d. All of the above can happen.

ANSWER KEY

Quiz 1

1. b		**9.** a	
2. a		**10.** b	
3. b		**11.** d	
4. a		**12.** d	
5. c		**13.** c	
6. b		**14.** b	
7. a		**15.** b	
8. d			

Quiz 2

1. b		**9.** c	
2. a		**10.** c	
3. c		**11.** d	
4. d		**12.** b	
5. c		**13.** a	
6. a		**14.** c	
7. b		**15.** a	
8. a			

CHAPTER 10: GENE EXPRESSION AND REGULATION

OUTLINE

Section 10.1 How Are Genes and Proteins Related?

- The experiments of Beadle and Tatum demonstrated that one gene encodes the information for the synthesis of one enzyme (**Figure 10-1**).
- **Ribonucleic acid** (**RNA**) carries the information from DNA to cytoplasmic ribosomes for protein synthesis. RNA is structurally different from DNA in that it is normally single stranded, has the sugar ribose, and replaces the use of the nitrogen-containing base thymine (T) with uracil (U) (**Table 10-1**).
- There are three forms of RNA. **Messenger RNA** (**mRNA**) carries the gene code from DNA to ribosomes. **Ribosomal RNA** (**rRNA**) combines with proteins to make ribsomes. **Transfer RNA** (**tRNA**) carries amino acids to ribosomes (**Figure 10-2**, **Table 10-2**).
- Protein synthesis is a two-step process. Within the nucleus, the DNA gene message is copied as mRNA by a process called **transcription** (RNA synthesis). The mRNA then binds to a ribosome where tRNA base pairs to mRNA, building the amino acid chain of a protein (**Figure 10-3**).
- The bases of mRNA are read as triplets (**codons**), each coding for the assembly of a single amino acid. The mRNA code for a protein begins with a **start codon** (AUG) and ends with **stop codons** (UAG, UAA, or UGA). The different combinations of bases in a codon code for the assembly of up to 20 amino acids (**Table 10-3**), but each codon specifies only one amino acid.

Section 10.2 How Is Information in a Gene Transcribed into RNA?

- Transcription consists of three steps: (1) initiation, (2) elongation, and (3) termination (**Figure 10-4**).
- **Initiation** occurs when transcription factors initiate the binding of the **promoter** region of the gene to **RNA polymerase**.
- **Elongation** occurs when RNA polymerase travels down the template strand of DNA, assembling RNA nucleotides into a single mRNA strand.
- Elongation continues until a **termination signal** DNA sequence is reached, causing RNA polymerase to release the assembled mRNA molecule and detach from DNA.

Section 10.3 How Is the Base Sequence of a Messenger RNA Molecule Translated into Protein?

- In prokaryotic cells, every nucleotide of a protein-coding gene codes for amino acids, while in eukaryotic cells these genes consist of **introns** and **exons**. Exons are regions that code for amino acids while introns do not, so the introns must be excised from the original RNA transcript before translation can commence (**Figures 10-6** and **10-7**).
- In eukaryotes, mRNA carries the genetic information to ribosomes in the cytoplasm where amino acid assembly occurs.
- Translation consists of three steps: (1) initiation, (2) elongation, and (3) termination (**Figure 10-8**).
- **Initiation** occurs when large and small ribosomal subunits come together at the first AUG codon of mRNA to form an initiation complex.
- **Elongation** occurs when tRNA molecules deliver amino acids to the mRNA molecule. The specific amino acid delivered depends on the base pairing between the tRNA **anticodon** and the mRNA codon. Two tRNA molecules can bind to the ribosome at the same time, with the large subunit causing the formation of a peptide bond between the amino acids. One tRNA then detaches from the ribosome, and the ribosome moves over one codon. A new tRNA then binds to the complementary mRNA codon, the amino acid is bonded to the growing amino acid chain, and the process repeats.
- **Termination** occurs when the ribosome encounters a stop codon, causing it to release the amino acid chain and then disassemble.

Section 10.4 How Do Mutations in DNA Affect the Function of Genes?

- **Mutations** are changes in the sequence of bases in DNA. Mutations can occur from normal errors in base pairing, as well as from environmental conditions that can cause damage to DNA (such as UV rays in sunlight).
- **Nucleotide substitutions** (**point mutations**) occur when the correct nucleotide in a pair is replaced instead of the incorrect one. **Insertion mutations** occur when one or more nucleotide pairs are inserted into the DNA double helix, while **deletion mutations** occur when pairs are removed from the DNA double helix (**Table 10-4**).
- **Inversions** are mutations that occur when chromosome pieces are cut out, turned around, and reinserted. **Translocations** are mutations that occur when a chunk of DNA is removed from a chromosome and moved to another one.
- Most mutations are harmful or neutral, but some can be beneficial and favored by natural selection, depending on the environment.

Section 10.5 How Are Genes Regulated?

- Gene expression occurs when it is transcribed and translated, forming a protein that causes an effect within the cell. The specific gene that is expressed is regulated by cell function, developmental stage, and the environment.
- Prokaryotic DNA is organized as packages (**operons**) in which genes for related functions lie close to each other (**Figure 10–10a**). Operons are regulated as units so functionally related proteins can be synthesized and expressed simultaneously. An example of this prokaryotic gene regulation mechanism is illustrated by the **lactose operon** of *E. coli* (**Figure 10–10b, c, d**).
- Eukaryotic gene expression is regulated at many steps: (1) Cells can control the frequency of individual gene transcription. (2) One gene may be used to produce different mRNAs and proteins. (3) Cells may control mRNA stability, as well as its translation. (4) Proteins may require modification before they are functional. (5) The lifespan of a protein can be regulated (**Figure 10–11**).
- Eukaryotic cells may regulate the transcription of genes, regions of chromosomes, or entire chromosomes by several mechanisms. Individual gene transcription can be altered by changing which transcription factors are present, thus affecting the ability of the promoter region to initiate transcription. Chromosomal regions that are condensed are not typically transcribed because they are inaccessible to RNA polymerase. Sometimes, entire chromosomes may remain condensed (**Figure 10–13**).

LEARNING OBJECTIVES

Section 10.1 How Are Genes and Proteins Related?

- Explain how Beadle and Tatum used a metabolic pathway to determine the one gene–one enzyme hypothesis.
- Discuss the structural and functional differences between DNA and RNA.
- Describe the overall processes of transcription and translation.
- Explain how transcription and translation are related.
- Explain how the genetic code bridges the gap between nucleic acids and amino acids.

Section 10.2 How Is Information in a Gene Transcribed into RNA?

- Explain the three stages of transcription.
- Discuss the similarities and differences between transcription and DNA replication.

Section 10.3 How Is the Base Sequence of a Messenger RNA Molecule Translated into Protein?

- Describe the differences between prokaryotic and eukaryotic transcription and translation.
- Discuss the differences between introns and exons.
- Discuss how different proteins can result from the same eukaryotic segment of DNA.
- Explain the structure and function of each ribosomal subunit.
- Discuss how the structure of tRNA provides for the insertion of the correct amino acid.
- Describe the order of assembly of the translation initiation complex.
- Describe the process of elongation, including peptide bond formation.
- Explain how stop codons terminate translation.
- Discuss how complementary base pairing maintains continuity from DNA to polypeptide synthesis.

Section 10.4 How Do Mutations in DNA Affect the Function of Genes?

- Explain why frame shift mutations are so dangerous.
- Describe the five possible outcomes resulting from a point mutation.
- Describe the effects of mutations in eggs and sperm on an organism.

Section 10.5 How Are Genes Regulated?

- Explain the necessity for gene regulation.
- Describe the structure and function of prokaryotic operons.
- Explain induction in prokaryotes using the lactose operon.
- Describe the five levels of control exercised by eukaryotic cells.
- Identify and explain the role of transcription factors.
- Discuss how chromosomal condensation prevents transcription.

QUIZ 1

1. RNA complementary to DNA is produced via
   _____.
   a. replication
   b. transcription
   c. translation
   d. protein synthesis

2. Imagine that a probe sent to Mars brings back a sample that contains a very primitive life-form that appears similar to bacteria. Scientists are able to revive it and begin to grow it in culture. Much to their amazement, they discover that the organism has DNA and that the DNA encodes proteins. However, the DNA of these Martian microbes contains only two nucleotides, and these nucleotides contain bases that are not present in the DNA of organisms on Earth. If the Martian microbe uses triplet codons, what is the maximum number of different amino acids that it can have in its proteins? Explain.
   a. 9, because $3^2 = 3 \times 3 = 9$
   b. 16, because $4^2 = 4 \times 4 = 16$
   c. 8, because $2^3 = 2 \times 2 \times 2 = 8$
   d. 7, because there are 8 possible codons ($2^3 = 2 \times 2 \times 2 = 8$) but at least one of the codons must be a stop.

3. Which of the following is an accurate statement concerning the differences between DNA and RNA?
   a. RNA is usually double stranded, but DNA is usually single stranded.
   b. RNA has the sugar deoxyribose, but DNA has the sugar ribose.
   c. RNA contains three different nucleotides, but DNA contains four different nucleotides.
   d. RNA lacks the base thymine (which is found in DNA) and has uracil instead.

4. What does mRNA carry from the nucleus?
   a. ribosomes
   b. information
   c. amino acids
   d. tRNA

5. Unripe black walnuts contain a compound, juglone, that inhibits RNA polymerase. With which process would juglone most likely interfere?
   a. mutation rate
   b. DNA replication
   c. transcription
   d. translation

6. All of the following are steps in transcription EXCEPT
   a. termination.
   b. elongation.
   c. initiation.
   d. recognition of the promoter by RNA polymerase.
   e. joining of short fragments of RNA by ligase.

7. After an RNA polymerase has completed transcription, the enzyme
   a. is degraded.
   b. joins with another RNA polymerase to carry on transcription.
   c. begins transcribing the next gene on the chromosome.
   d. is free to bind to another promoter and begin transcription.
   e. will remain bound to the RNA.

8. The information to synthesize proteins is carried to the ribosome by
   a. tRNA.
   b. mRNA.
   c. rRNA.
   d. template DNA.
   e. DNA of the gene.

9. Which molecule is responsible for bringing the correct amino acid to the ribosome at the correct time?
   a. rRNA
   b. mRNA
   c. small subunit
   d. large subunit
   e. tRNA

10. After translation is completed, the ribosome
    a. stays intact and finds another AUG codon to begin translation.
    b. moves back to the first AUG codon of the mRNA to begin again.
    c. joins with other ribosomes to continue translation.
    d. breaks into small and large ribosomal subunits.
    e. will remain bound to the mRNA.

11. The capacity for forming the peptide bond between two amino acids resides in the
    a. catalytic site of the large subunit.
    b. tRNA.
    c. proteins involved in termination of translation.
    d. the amino acids themselves.

12. Some people have eyes of two different colors. What is a possible explanation for this trait?
    a. A mutation occurred in the sperm that produced this individual's embryo.
    b. A mutation occurred in the egg that produced this individual's embryo.
    c. A chemical toxin inhibited production of pigment in one eye but not the other.
    d. During early stages of development, a mutation occurred in the cell that developed into one of the eyes, but not other cells in the embryo.

13. During DNA replication, a mistake was made in which an A was changed to a G. This kind of mutation is called a(n) _____.
    a. point mutation
    b. insertion mutation

c. deletion mutation
d. neutral mutation

14. The cells in your skin have a different shape and different function from the cells in your liver because the two types of cells have different _____.
    a. DNA
    b. proteins
    c. lipids
    d. carbohydrates

15. In mammals, males have one X chromosome and one Y chromosome and females have two X chromosomes. How is the expression of genes on the X chromosome regulated so that there is equal expression of genes on the X chromosome in males and females?
    a. One X chromosome in females is inactivated so that females have only a single X chromosome capable of transcription.
    b. The genes on the X chromosome in males are transcribed twice as fast as in females.
    c. All of the X chromosomes are inactivated so that no genes are expressed from the X chromosome in either males or females.
    d. The Y chromosome contains balancing genes that help raise the levels of mRNA produced by the X chromosome in males.

QUIZ 2

1. Which of the following statements about the functions of RNA is correct?
    a. The information for protein synthesis is carried by tRNA.
    b. rRNA is an intermediate in the synthesis of mRNA.
    c. rRNA is an important component of ribosomes.
    d. Translation requires tRNA and mRNA, but not rRNA.

2. The information to synthesize all of the following EXCEPT _____ is stored in the DNA.
    a. organelles
    b. messenger RNA
    c. proteins
    d. ribosomal RNA
    e. DNA

3. The process that uses the genetic information carried by mRNA to specify the sequence of amino acids in a protein is called
    a. replication.
    b. translation.

c. transcription.
d. synthesis.
e. elongation.

4. One way transcription differs from DNA replication is
    a. only DNA replication involves base pairing to determine the next nucleotide in the chain.
    b. transcription requires only one enzyme, DNA replication several.
    c. transcription produces a completely new synthesized strand; DNA replication produces a completely new synthesized double helix.
    d. transcription proceeds continuously; DNA replication proceeds discontinuously.
    e. there are no real differences.

5. What determines which of the two DNA strands will serve as the template?
    a. Only one of the strands can be transcribed.
    b. the sequence of the bases in the RNA after transcription determines which will serve as the template.

c. The RNA polymerase can transcribe only one strand of DNA.

d. Only one strand will unwind.

e. The orientation of the promoter and other regulatory sequences determines which will serve as the template.

6. The next base to be added to the RNA strand is determined by

a. the previous base.

b. the order of the backbone in the RNA.

c. base pairing between the two DNA strands.

d. base pairing between the template strand and the RNA nucleotides.

7. If you were to create a drug that recognized and bound irreversibly to the promoter region of a specific gene, thereby blocking it, what effect would you expect that drug to have?

a. There would be no effect on the gene.

b. Only transcription of that gene would halt.

c. Only translation of the gene would halt.

d. Both transcription and, eventually, translation of the gene would halt.

8. What is the relationship between codons and amino acids?

a. For every amino acid, there is a unique codon.

b. All codons code for amino acids.

c. There are more amino acids than codons.

d. There are more codons than amino acids.

9. Which of the following is the first step in translation?

a. Bases of the tRNA anticodon bind with the bases of the mRNA codon.

b. A peptide bond forms between amino acids attached to the adjacent tRNAs on the ribosome.

c. The ribosomal subunits are disassembled.

d. Stop codons on the mRNA bind to special proteins rather than to tRNA molecules.

10. The flow of genetic information in cells depends on specific base pairing between nucleotides. Which of the following correctly matches the type of base pairing with the process of translation?

a. RNA base-pairs with DNA.

b. rRNA base-pairs with mRNA.

c. tRNA base-pairs with rRNA.

d. tRNA base-pairs with mRNA.

11. Imagine that a codon in the template strand of a gene has the sequence TAC. What sequence of the anticodon would decode this codon? Explain your answer.

a. ATG, because the anticodon is complementary to the template strand

b. AUG, because the anticodon is complementary to the template strand

c. UAC, because the anticodon has the same sequence as the template strand (but it has U instead of T)

d. TAC, because the anticodon has the same sequence as the template strand

12. Which kind of point mutation would have the most dramatic effect on the protein coded for by that gene?

a. a base substitution

b. a base insertion near the beginning of the coding sequence

c. a base insertion near the end of the coding sequence

d. deletion of three bases near the start of the coding sequence

13. The genes that are expressed in a cell depend on the cell's _____.

a. environment

b. history

c. function

d. all of the above

14. Which of the following is NOT a step at which gene expression can be regulated in eukaryotic cells?

a. rate of transcription

b. rate of translation

c. rate of DNA replication

d. rate of enzyme activity

15. Certain genes, sometimes called *housekeeping genes,* are expressed in all cells in your body. Other genes are expressed only in certain specialized cells. Which of the following genes is likely to be a housekeeping gene?

a. hemoglobin

b. milk proteins

c. ribosomal proteins

d. insulin

Quiz 1

1. b
2. d
3. d
4. b
5. c
6. e
7. d
8. b

9. e
10. d
11. a
12. d
13. a
14. b
15. a

Quiz 2

1. c
2. a
3. b
4. b
5. e
6. d
7. d
8. d

9. a
10. d
11. c
12. b
13. d
14. c
15. c

CHAPTER 11: THE CONTINUITY OF LIFE: CELLULAR REPRODUCTION

OUTLINE

Section 11.1 What Is the Role of Cellular Reproduction in the Lives of Individual Cells and Entire Organisms?

- The **cell cycle** is the sequence of activities that occur from one cell division to the next. Cell division enables asexual reproduction, during which offspring are formed from a single parent (**Figure 11-1**).
- The prokaryotic cell cycle consists of growth and DNA replication, followed by **binary fission**. Binary fission results in the formation of two cells with identical DNA (**Figure 11-2**).
- The eukaryotic cell cycle consists of interphase and cell division. During **interphase**, the cell grows and replicates its DNA over three phases (G_1, S, and G_2). In the G_1 phase, the cell grows and acquires materials needed for cell division. DNA synthesis occurs during the S phase. Growth is completed during the G_2 phase. Instead of entering the S phase, a cell may instead enter the G_0 phase and **differentiate** (**Figure 11-3**).
- During **cell division**, a cell may duplicate itself asexually by **mitosis** (nuclear division) followed by **cytokinesis**. Mitosis can be used to maintain and repair tissues. Alternatively, cells of the ovaries and testes may duplicate, forming genetically unique **gametes** by **meiosis** in anticipation of **sexual reproduction**.

Section 11.2 How Is DNA in Eukaryotic Cells Organized into Chromosomes?

- Eukaryotic **chromosomes** consist of linear DNA molecules bound to proteins, which allows otherwise long molecules to be wrapped as tightly coiled packages (**Figure 11-4**). Chromosomes are composed of **telomeres** (regions that stabilize a chromosome) and a **centromere** (which joins replicated DNA helices) (**Figure 11-5**).
- **Duplicated chromosomes** are composed of identical **sister chromatids** that can separate to form **independent chromosomes** (**Figures 11-6, 11-7,** and **11-8**).
- Eukaryotic chromosomes typically occur in **homologous pairs** with similar genetic information. Cells with these pairs (most of them) are called **diploid**, while those with only one of each chromosome pair are called **haploid**. Cell chromosomes have both **autosomes** and **sex chromosomes**.

Section 11.3 How Do Cells Reproduce by Mitotic Cell Division?

- During **interphase**, chromosomes are duplicated, forming sister chromatids. These then enter the four stages of mitosis: prophase, metaphase, anaphase, and telophase. Cytokinesis usually follows telophase (**Figure 11-10**).
- During **prophase**, chromosomes condense and their **kinetochores** attach to spindle microtubules. The nuclear membrane breaks down
- During **metaphase**, sister chromatids move to the equator of the cell.
- During **anaphase**, sister chromatids separate and are pulled along their spindle microtubules to the cell poles.
- During **telophase**, the chromosomes uncoil and nuclear envelopes reform around each nucleus.
- During **cytokinesis**, the cytoplasm is divided into nearly equal halves by either contracting microfilaments (animal cells) or the fusion of vesicles along the center of the cell (plant cells) (**Figures 11-11** and **11-12**).

Section 11.4 How Is the Cell Cycle Controlled?

- Complex protein interactions drive the cell cycle. The major **checkpoints** by which the cell cycle is regulated occur (1) between G_1 and S, (2) between G_2 and mitosis, and (3) between metaphase and anaphase (**Figures 11-13, 11-14,** and **11-15**).

Section 11.5 Why Do So Many Organisms Reproduce Sexually?

- Mutations are the origin of genetic variability, forming **alleles** that can produce differences in structure and function (**Figure 11-16**).

- Sexual reproduction can combine parental alleles in unique ways within offspring, creating variation that can improve the survival chances of an organism, as well as its ability to reproduce.

Section 11.6 How Does Meiotic Cell Division Produce Haploid Cells?

- Meiosis produces haploid cells by separating **homologous chromosomes** so that each cell contains only one homologue (**Figures 11-17, 11-18**, and **11-19**).
- Chromosomes are duplicated during **interphase**, and then the cell undergoes two specialized cell divisions (**meiosis I** and **meiosis II**) to produce four haploid cells (**Figure 11-21**).
- **Meiosis I**: During **prophase I**, homologous duplicated chromosomes pair up and **crossing over** occurs. During **metaphase I**, paired homologous chromosomes line up along the cell equator. During **anaphase I**, paired homologous chromosomes separate and are pulled to opposite poles of the cell. New nuclei form during **telophase I**. Since each daughter cell only has one of each pair of homologues, it is considered haploid.
- **Meiosis II**: Both haploid daughter nuclei undergo divisions similar to mitosis. Nuclear membranes break down during **prophase II**. During **metaphase II**, the two chromatids of each chromosome move to the cell equator. During **anaphase II**, the two chromatids separate, each moving to opposite cell poles. During **telophase II**, nuclear membranes re-form, producing four haploid nuclei. **Cytokinesis** typically occurs after (or during) telophase II, producing four haploid cells.

Section 11.7 When Do Mitotic and Meiotic Cell Division Occur in the Life Cycles of Eukaryotes?

- Eukaryotic life cycles have three parts: (1) fertilization causes the fusion of gametes, forming a diploid cell; (2) meiotic cell division occurs, creating haploid cells; and (3) the growth of multicellular bodies or asexual reproduction occurs (**Figure 11-25**).
- The proportion of time spent in these stages is dependent on the life cycle of the particular species. **Haploid life cycles** predominantly consist of haploid cells (**Figure 11-26**). Diploid life cycles predominantly consist of diploid cells (**Figure 11-27**). **Alternation of generations life cycles** have both diploid and haploid multicellular stages (**Figure 11-28**).

Section 11.8 How Do Meiosis and Sexual Reproduction Produce Genetic Variability?

- Genetic variability occurs through the shuffling of homologous maternal and paternal chromosomes during metaphase I (**Figure 11-29** and **11-30**), crossing over during prophase I, and the fusion of gametes during fertilization.

LEARNING OBJECTIVES

Section 11.1 What Is the Role of Cellular Reproduction in the Lives of Individual Cells and Entire Organisms?

- Explain the difference between sexual reproduction and asexual reproduction.
- Describe the steps of binary fission in prokaryotes.
- Discuss the activities occurring during interphase (G_1, S and G_2).
- Explain the multiple functions of mitosis.
- Explain the similarities and differences between mitosis and meiosis.

Section 11.2 How Is DNA in Eukaryotic Cells Organized into Chromosomes?

- Describe the structure of a chromosome, including the centromeres and telomeres.
- Explain the difference between the terms *chromosome* and *chromatids*.
- Describe the characteristics of autosomes and sex chromosomes.
- Discuss the locations and functions of diploid and haploid cells.

Section 11.3 How Do Cells Reproduce by Mitotic Cell Division?

- Describe the activities of the four stages of mitosis.
- Explain how daughter cells are provided with an exact copy of chromosomes from the parent cell.

Section 11.4 How Is the Cell Cycle Controlled?

- Describe the cell cycle checkpoints.
- Explain the cascade effect initiated by growth factors that trigger cell division.
- Explain the relationship between the genes Rb and P53 and cancer.

Section 11.5 Why Do So Many Organisms Reproduce Sexually?

- Explain the relationship of an allele to a gene.
- Explain how sexual reproduction increases variability in a species.

Section 11.6 How Does Meiotic Cell Division Produce Haploid Cells?

- Explain the chromosome number differences resulting from mitosis and meiosis.
- Explain the function of meiosis.
- Discuss the differences between meiosis and mitosis.
- Explain the necessity of meiosis for sexually reproducing organisms.
- Discuss the two factors that increase the genetic variation between gametes.

Section 11.7 When Do Mitotic and Meiotic Cell Division Occur in the Life Cycles of Eukaryotes?

- Describe the three different diploid/haploid life cycles.

Section 11.8 How Do Meiosis and Sexual Reproduction Produce Genetic Variability?

- Describe all factors of diploid organisms that contribute to genetic variability within a species.

QUIZ 1

1. The copying of chromosomes occurs during
   a. G_1.
   b. S phase
   c. G_2.
   d. mitosis.

2. Which of the following statements about the chromosomes of prokaryotic and eukaryotic cells is TRUE?
   a. Both prokaryotic and eukaryotic cells have multiple chromosomes.
   b. The chromosome of a prokaryotic cell is a circular DNA double helix, but the chromosomes of eukaryotic cells are linear DNA double helices.
   c. The chromosome of prokaryotic cells is present in their nuclei, but the chromosomes of eukaryotic cells are in the cytoplasm.
   d. Chromosomes of eukaryotic cells are attached to the plasma membrane, but the chromosome of prokaryotic cells floats free in the cytoplasm.

3. A duplicated chromosome contains _____.
   a. two DNA double helices
   b. two sister chromatids

   c. four strands of DNA
   d. all of the above
   e. none of the above

4. Imagine that you are looking at a eukaryotic cell in a microscope. When you examine the cell's nucleus, you see that the chromatin is spread uniformly through the nucleus—you cannot see chromosomes. Has the cell's DNA been replicated yet? Explain.
   a. The DNA has been replicated because DNA replication occurs during interphase.
   b. The DNA has not replicated because DNA replication occurs after chromosome condensation.
   c. You cannot tell if the DNA has replicated unless the DNA is condensed.

5. A clone is _____.
   a. an unnatural creature fabricated by unscrupulous scientists
   b. any cell or organism that is genetically identical to another
   c. an exact duplicate of an organism
   d. a sexually reproduced organism

6. During prophase of mitosis
  a. the chromatids start to separate.
  b. chromosomes decondense.
  c. the nucleolus re-forms.
  d. centrioles are produced.
  e. the spindle microtubules attach to the chromosomes.

7. A chemical that is potentially useful for treating cancer would
  a. prevent recombination.
  b. prevent DNA synthesis.
  c. inhibit pairing of homologous chromosomes.
  d. induce mutations.

8. Alternate forms of a particular gene are called _____; they arise as a result of _____.
  a. alleles, meiosis
  b. mutations, mitosis
  c. alleles, mutation
  d. clones, sexual reproduction

9. Asexual reproduction differs from sexual reproduction in that asexual reproduction
  a. produces greater genetic variation than does sexual reproduction.
  b. allows genes to be shuffled more readily than does sexual reproduction.
  c. can occur more quickly than sexual reproduction.
  d. cannot contribute to the growth of multicellular organisms, whereas sexual reproduction can.

10. *Meiosis* comes from a Greek word that means "to decrease." What decreases during the process of meiosis?
  a. size of chromosomes
  b. number of cells
  c. length of the DNA double helices
  d. number of chromosomes

11. During the process of meiosis, DNA is replicated _____, followed by _____ nuclear division(s).
  a. twice, two
  b. twice, one
  c. once, two
  d. once, one

12. During meiosis I, _____ separate; during meiosis II, _____ separate.
  a. homologous chromosomes, sister chromatids
  b. sister chromatids, homologous chromosomes
  c. sister cells, gametes
  d. DNA double helices, DNA double helices

13. Imagine that you are looking at a eukaryotic cell in the microscope. When you examine the cell, you see that the nucleus is not present and that chromosomes are condensed and lined up independently in single file in the center of the cell. You might be observing metaphase of _____.
  a. mitosis
  b. meiosis I
  c. meiosis II
  d. mitosis or meiosis I
  e. mitosis or meiosis II

14. Genetic recombination (crossing over) produces _____.
  a. new alleles
  b. mutations
  c. new combinations of alleles
  d. longer chromosomes

15. Sexual reproduction produces genetic diversity by _____.
  a. creating new combinations of homologous chromosomes
  b. creating new combinations of alleles
  c. fusing gametes to form the diploid organism
  d. all of the above

QUIZ 2

1. When a cell divides, what must it pass on to its offspring?
  a. a complete set of chromosomes
  b. a complete set of messenger mRNA molecules so that cells can express every gene
  c. cytoplasmic components needed for transcription and translation
  d. all of the above
  e. (a) and (c)

2. The term *haploid* refers to
  a. chromosomes that contain the same genes.
  b. cells that contain a pair of each type of chromosome.
  c. a complete set of chromosomes from a single cell that have been stained for microscopic examination.
  d. cells that contain only one of each type of chromosome.

3. Immediately after replication, the two DNA molecules
  a. remain attached at their centromeres.
  b. head straight into the mitosis phase.
  c. are moved to the equator of the cell.
  d. are held together by the kinetochore.

4. During anaphase, the cell becomes more ovoid because
  a. microtubules push the poles of the cell apart.
  b. the microfilament ring around the equator contracts.
  c. there are excess amounts of cytoplasm.
  d. the chromosomes line up at the equator.
  e. the centrioles migrate to opposite ends of the cell.

5. Nuclear envelopes form during
   a. anaphase.
   b. prophase.
   c. metaphase.
   d. telophase.

6. Cytokinesis following mitosis differs in animal and plant cells because
   a. plant cells have chloroplasts.
   b. plant cells have cell walls.
   c. animal cells have mitochondria.
   d. animal cells have more cytoplasm.

7. An advantage of sexual reproduction is that it
   a. promotes genetic variability, thereby increasing the probability that an individual with new combinations of favorable traits may arise.
   b. ensures that individuals will inherit the most desirable genes from both parents.
   c. ensures that offspring are as similar as possible to their parents.
   d. none of the above; it is an evolutionary relic of an earlier stage.

8. What is the ultimate source of genetic variability in organisms?
   a. sexual reproduction
   b. DNA replication
   c. homologous chromosomes
   d. mutations in DNA

9. Mating a male donkey to a female horse produces mules. Horses have 64 chromosomes and donkeys have 62 chromosomes. How many chromosomes do mules have? Why are they sterile?
   a. The mule has 126 chromosomes and is sterile because 126 chromosomes are too many to go through meiosis.
   b. The mule has 64 chromosomes, as does its mother. It is sterile as a result of mutations that prevent sperm production.
   c. The mule has 63 chromosomes and is sterile because the chromosomes cannot pair properly at metaphase of meiosis I.
   d. The mule has 63 chromosomes and is sterile because the chromosomes cannot pair properly at metaphase of meiosis II.

10. If a diploid cell replicates its DNA so that it now contains an amount of DNA equal to $4n$, how does a haploid gamete get a $1n$ number of chromosomes and a $1n$ amount of DNA?
   a. two meiotic divisions and four daughter cells produced in meiotic cell division
   b. one meiotic division and two daughter cells produced in meiotic cell division
   c. one meiotic division and four daughter cells produced in meiotic cell division
   d. none of the above

11. The random alignment of homologues at the equator during metaphase I is called
   a. crossing over.
   b. chiasmata.
   c. random alignment.
   d. independent assortment.

12. Cells that have only one of each type of chromosome, either the maternal or paternal homologue, are considered haploid. After which step in meiosis is this first seen?
   a. metaphase I
   b. anaphase I
   c. telophase I
   d. cytokinesis I

13. After meiosis, the resulting haploid cells will
   a. have the same genetic material as the parental cell.
   b. have the same number of chromosomes as the parent cell.
   c. have the same amount of cytoplasm as the parental cell.
   d. have one member from each pair of homologous chromosomes.

14. Mitotic cell division allows for all of the following to occur EXCEPT
   a. bodily repair.
   b. tissue and organ regeneration.
   c. production of sperm and eggs.
   d. growth of an individual.
   e. replacement of dead cells.

15. Which event(s) is (are) responsible for the genetic variability seen in meiosis?
   a. The direction in which a parental chromosome faces during metaphase I is random.
   b. Homologous chromosomes exchange DNA with one another.
   c. Homologous chromosomes exchange RNA with one another.
   d. (a) and (b) are correct.

ANSWER KEY

Quiz 1

1. b
2. b
3. d
4. c
5. b
6. e
7. b
8. c

9. c
10. d
11. c
12. a
13. d
14. c
15. d

Quiz 2

1. e
2. d
3. a
4. a
5. d
6. b
7. a
8. d

9. c
10. a
11. d
12. d
13. e
14. c
15. d

CHAPTER 12: PATTERNS OF INHERITANCE

OUTLINE

Section 12.1 What Is the Physical Basis of Inheritance?

- **Inheritance** occurs when characteristics of individuals are passed to their offspring. These characteristics are passed on in the form of **genes**, which are located on **chromosomes**.
- Diploid cells contain pairs of homologous chromosomes that carry the same set of genes. Each gene is located at the same **locus** on its chromosome.
- Differences in genes at the same locus produce different **alleles** of that gene (**Figure 12-1**). Homologous chromosomes with the same allele at a locus are called **homozygous**, those with different alleles are called **heterozygous**.

Section 12.2 How Did Gregor Mendel Lay the Foundations for Modern Genetics?

- Gregor Mendel discovered many of the principles of inheritance. He did this by being the first geneticist to choose the right organism, to design and perform the experiment properly, and to analyze the data correctly.

Section 12.3 How Are Single Traits Inherited?

- Traits inheritance depends on the allele composition of the parents. Each parent contributes one copy of every gene during gamete formation (**law of segregation**), resulting in the inheritance of a pair of alleles for each gene by the offspring.
- **Dominant alleles** mask the expression of **recessive alleles**, resulting in offspring with the same **phenotype** but different **genotypes**. For example, **homozygous dominant** organisms have the same phenotype as those who are **heterozygous**.
- Each allele separates randomly during meiosis, allowing for the prediction of inherited traits according to the laws of probability. This can be done through the use of the **Punnett square method** (**Figure 12-11a**).

Section 12.4 How Are Multiple Traits Inherited?

- If the genes for multiple traits are on separate chromosomes, their inheritance follows the **law of independent assortment**. This law states that alleles of one gene may be distributed to gametes independently of the alleles of other genes and is due to the random arrangement of homologous pairs during metaphase I (**Figure 12-15**).
- As with the inheritance of single traits, the inheritance of multiple traits can be predicted according to the laws of probability using the Punnett square method (**Figure 12-14**).

Section 12.5 How Are Genes Located on the Same Chromosome Inherited?

- Genes located on the same chromosome are **linked** and tend to be inherited together (**Figure 12-16**). However, some linked genes may be inherited separately as the result of **crossing over (genetic recombination)** that occurs during prophase I (**Figures 12-17** to **12-20**).

Section 12.6 How Is Sex Determined, and How Are Sex-Linked Genes Inherited?

- The sex of many animals is determined by sex chromosomes, often represented as X and Y. All other chromosomes in a cell are **autosomes**. Most females have two X chromosomes, as opposed to one X and one Y in males. Sex is determined by the Y chromosome carried in the sperm of the male (**Figure 12-22**).
- The Y chromosome is small and carries few genes compared to the X chromosome. Therefore, recessive traits that are carried on the X chromosome are more likely to be phenotypically expressed in males (**Figure 12-23**).

Section 12.7 Do the Mendelian Rules of Inheritance Apply to All Traits?

- Not all traits follow Mendelian inheritance patterns. **Incomplete dominance** occurs when heterozygotes are an intermediate between the homozygous phenotypes (**Figure 12-24**). **Codominance** occurs when different alleles at one locus both contribute to the phenotype (e.g., blood types). **Polygenic inheritance** occurs when some phenotypic traits are determined by several different genes at different loci (**Figure 12–25**). **Pleiotrophy** occurs when genes have multiple effects on phenotype (e.g., the SRY gene). The **environment** can affect most, if not all, types of phenotypic expression (**Figure 12–26**).

Section 12.8 How Are Human Genetic Disorders Investigated?

- Human genetic disorders can be investigated by analysis of the family history of genetic crosses (**pedigrees**) and use of molecular genetic technologies (**Figure 12–27**).

Section 12.9 How Are Human Disorders Caused by Single Genes Inherited?

- Some genetic disorders, such as albinism and sickle-cell anemia, are caused by recessive alleles (**Figures 12–28** and **12–29**). The disorder is only phenotypically expressed in homozygotes. Heterozygotes do not express the disorder but do carry the recessive allele.
- Some genetic disorders, such as Huntington's disease, are caused by dominant alleles. Thus, inheriting only one dominant allele is sufficient for the offspring to express the disorder phenotypically.
- Some genetic disorders, such as hemophilia and red-green color blindness, are sex-linked. The allele carrying the disorder is carried on the X chromosome; thus, males have a much greater chance of being affected because they lack a second X chromosome (**Figures 12–30** and **12–31**).

Section 12.10 How Do Errors in Chromosome Number Affect Humans?

- Errors in chromosome number occur through **nondisjunction** of chromosomes during meiosis (**Figure 12–32**).
- Some disorders, such as **Turner syndrome**, **trisomy X**, **Klinefelter syndrome**, and **Jacob syndrome**, are caused by an abnormal number of sex chromosomes. These disorders result in distinctive physical characteristics derived from the sex chromosomes that are affected.
- Some disorders, such as trisomy 21 (**Down syndrome**), are caused by abnormal numbers of autosomes. These disorders usually result in spontaneous abortions, but for those fetuses that survive, severe mental and physical deficiencies occur. The likelihood of these disorders increases with the age of the mother and, to a lesser degree, the father (**Figure 12–34**).

LEARNING OBJECTIVES

Section 12.1 What Is the Physical Basis of Inheritance?

- Describe the relationship among DNA, genes, and chromosomes.
- Explain the terms *alleles* and *loci* with respect to genes.
- Explain the difference between homozygous and heterozygous.

Section 12.2 How Did Gregor Mendel Lay the Foundations for Modern Genetics?

- Describe the attributes of the garden pea that allowed Mendel's experiments to work.

Section 12.3 How Are Single Traits Inherited?

- Describe Mendel's experimental design from the parental generation through the F_2 generation.
- Describe the relationship between dominant and recessive alleles.
- Describe the genotype for the parental, F_1, and F_2 generation plants for flower color.
- Explain how a purple flower plant can self-fertilize and produce white flower plants.
- Explain the relationship between genotype and phenotype.
- Describe the steps for completing a Punnett Square.
- Explain why a test cross requires the use of the recessive phenotype.

Section 12.4 How Are Multiple Traits Inherited?

- Explain the law of independent assortment citing a two-factor cross.

Section 12.5 How Are Genes Located on the Same Chromosome Inherited?

- Explain the relationship between genetic linkage and independent assortment of alleles.
- Explain how crossing over increases genetic diversity.

Section 12.6 How Is Sex Determined, and How Are Sex-Linked Genes Inherited?

- Explain how the inheritance and phenotypic expression of genes on the X chromosome differs by gender.

Section 12.7 Do the Mendelian Rules of Inheritance Apply to All Traits?

- Describe the difference between the heterozygous phenotype of a gene with complete dominance with that of one with incomplete dominance.
- Describe how multiple alleles arise for a given trait.
- Describe the difference in expression for alleles that are codominant with those that are incompletely dominant.
- Explain the variation of human height with respect to polygenic inheritance.
- Discuss the difference between pleiotropy and polygenic inheritance.
- Describe an instance in which the environment influences the expression of a phenotype.

Section 12.8 How Are Human Genetic Disorders Investigated?

- Explain how a pedigree can help determine whether a trait is dominant or recessive.

Section 12.9 How Are Human Disorders Caused by Single Genes Inherited?

- Explain the role of carriers in recessive genetic disorders.
- Explain how a mutant allele can be dominant to the normal allele.
- Discuss the difference in inheritance patterns for recessive and dominant genetic disorders.
- Explain the inheritance pattern by gender for sex-linked disorders.

Section 12.10 How Do Errors in Chromosome Number Affect Humans?

- Explain the role of nondisjunction in Turner's syndrome.
- Discuss the age-related increase in Down syndrome.

QUIZ 1

1. Alleles are
   a. specific physical locations of genes on a chromosome.
   b. variations of the same gene (i.e., similar nucleotide sequences on homologous chromosomes).
   c. homozygotes.
   d. heterozygotes.

2. In peas, plants that are true-breeding for a particular trait must be _____ for that trait.
   a. heterozygous
   b. diploids
   c. homozygous
   d. either homozygous or heterozygous

3. The appearance of an organism that results from its particular complement of genes is called its
   a. genotype.
   b. dominance.
   c. phenotype.
   d. inheritance.

4. In pea plants, yellow pods are recessive to green pods. If you see yellow pods, then the genotype of that plant must be _____ for pod color.
   a. heterozygous
   b. homozygous recessive
   c. homozygous dominant
   d. (a) and (b)
   e. (a) and (c)

5. In pea plants, tall plants are dominant over dwarf plants. A tall plant could be
   a. heterozygous.
   b. homozygous recessive.
   c. homozygous dominant.
   d. (a) and (b)
   e. (a) and (c)

6. Mendel's law of segregation concludes that
   a. all gametes formed by an organism will have the same allele.
   b. each individual has two alleles for a particular gene.
   c. genes are found at the same loci on homologous chromosomes.
   d. dominant alleles may mask the expression of recessive alleles.
   e. each gamete will contain only one allele from the parent's pair.

7. What does Mendel's law of independent assortment tell us about the behavior of genes during meiosis?
   a. Alleles of a particular gene will be distributed to gametes randomly, independent of other genes on different chromosomes.
   b. Alleles on the same chromosome will be distributed to different gametes independently.
   c. Independent assortment is synonymous with crossing over.
   d. All of the above are correct.

8. In pea plants, green seeds are recessive to yellow and wrinkled seeds are recessive to smooth seeds. Assume a plant produces green, wrinkled seed. Using S to designate seed shape and Y to designate seed color, this plant must have the following genotype:
   a. *ssYy*.
   b. *Ssyy*.
   c. *SsYy*.
   d. *ssYY*.
   e. *ssyy*.

9. When Mendel crossbred plants that differed in two traits, he found that the phenotypic ratio was
   a. 4:4:4:4.
   b. 9:3:3:1.
   c. 6:6:2:2.
   d. 7:7:1:1.
   e. The ratios varied, depending on the traits being tested.

10. Which of the following can account for a situation in which Mendel's law of independent assortment fails to hold?
   a. genes on the same chromosome
   b. pleiotropy
   c. self-fertilization
   d. genes have undergone recombination
   e. crossing over

11. Anne Boleyn, King Henry VIII's second wife, was beheaded because she did not provide him with a son as an heir. Explain why King Henry should have blamed himself and not his wife.
   a. All the sperm that males produce contain an X chromosome, so their genetic contribution to the child determines its sex.
   b. All the eggs that females produce contain an X chromosome, so their genetic contribution to the child does not determine its sex.
   c. The eggs that females produce contain either an X or a Y chromosome, so their genetic contribution to the child is unrelated to its sex.

12. A single gene capable of influencing multiple phenotypes within a single organism is said to be
   _____.
   a. codominant for that gene
   b. incompletely dominant for that gene
   c. polygenic for that gene
   d. pleiotropic for that gene

13. If a gene has alleles that are incompletely dominant, an individual that is heterozygous at this locus will express a phenotype that is _____.
   a. the same as organisms that are homozygous for the recessive allele
   b. the same as organisms that are homozygous for the dominant allele
   c. intermediate between organisms that are homozygous for the recessive allele and organisms that are homozygous for the dominant allele

14. Cystic fibrosis is a recessive trait. Imagine that your friend Roger has cystic fibrosis but that his parents do not. What do you know about Roger's alleles and those of his parents at the cystic fibrosis locus of their DNA?
   a. This information is insufficient to allow me to conclude anything about the cystic fibrosis alleles in the DNA of Roger's parents.
   b. This information is insufficient to allow me to conclude anything about the cystic fibrosis alleles in Roger's DNA.
   c. Roger is heterozygous and his parents are homozygous at the cystic fibrosis locus.
   d. Roger is homozygous and his parents are heterozygous at the cystic fibrosis locus.

15. What happens if a baby has only one X chromosome and no Y?
   a. Such a deficiency is lethal, so the baby would be stillborn.
   b. This baby would be a female with Turner syndrome.
   c. The baby would be a male with Turner syndrome.
   d. The baby would have Klinefelter syndrome.

QUIZ 2

1. Alleles are alternate forms of a gene. The alleles for the gene that determines blood type in humans are found at _____.
   a. different loci on homologous chromosomes
   b. different loci on the same chromosome
   c. the same locus on homologous chromosomes

2. Humans have about 35,000 genes. How many alleles (either the same or different) of EACH of these genes are present in your muscle cells, disregarding genes on the X and Y chromosomes?
   a. 1
   b. 2
   c. 23
   d. 46

3. Mendel was able to determine the dominant and recessive forms of seven traits in pea plants by
   a. crossing plants that differed in two traits.
   b. looking at the offspring from self-fertilization of true-breeding plants.
   c. a simple examination of the plants.
   d. expanding his research to include other plants.
   e. doing several crosses using plants that differed in one trait.

4. When Mendel conducted his experiments with purple- and white-flowered pea plants, he found that the F_1 generation did not contain any white-flowered plants. In the F_2 generation, the white-flowered plants were seen. Which of the following statements is TRUE?
   a. The purple phenotype is always present in every generation.
   b. The white-flower allele is dominant to the purple-flower allele.
   c. The white-flower phenotype shows up only in alternate generations.
   d. Many factors are involved in this "masking" phenomenon.
   e. The white-flower allele is recessive to the purple-flower allele.

5. The mating of an individual that is true-breeding for the dominant phenotype with an individual that is true-breeding for the recessive phenotype will
   a. produce no offspring.
   b. give rise to offspring in which expression of the recessive allele is masked.
   c. produce only male organisms.
   d. produce offspring exhibiting both the dominant and recessive phenotypes.
   e. produce a generation consisting of individuals exhibiting only the recessive phenotype.

6. In experiments using pea plants, two purple-flowered plants are cross-fertilized. Some of the F_1 generation plants have white flowers. In order for this to occur

   a. both parental plants must have been homozygous recessive.
   b. both parental plants must have been homozygous dominant.
   c. parental plants must both have been heterozygous.
   d. one of the parents must be heterozygous, while the other is homozygous dominant.

7. In pea plants, the white-flower trait is recessive to the purple-flower trait, and the dwarf trait is recessive to the tall trait. Assume a pea plant that is tall with white flowers. Using T to designate the plant size gene and P to designate flower color, this would require a genotype of
   a. *PPTt.*
   b. *ppTT* or *ppTt.*
   c. *pptt.*
   d. *pptt* or *ppTt.*
   e. *PpTt.*

8. In the mating between two individuals that are heterozygous for two traits
   a. the offspring will have approximately the same number of individuals showing dominant phenotypes as the recessive phenotypes.
   b. the offspring will have only the dominant phenotype for each trait.
   c. one-quarter of the offspring will exhibit one recessive phenotype and $\frac{1}{4}$ will exhibit the other recessive phenotype.
   d. there will be three times as many offspring with the recessive phenotype as with the dominant phenotype for one of the traits.
   e. the offspring will have only the recessive phenotype for each trait.

9. In the results of a dihybrid cross, an individual displays the recessive form for both traits. Which of the following statements describing the individual is true?
   a. This overall phenotype is the most abundant among the offspring.
   b. The individual must be heterozygous for both genes involved.
   c. The individual must be homozygous recessive for at least one of the two genes involved.
   d. The individual must have two recessive alleles for each of the genes in question.
   e. This individual will be weaker than the individual showing the dominant form for both traits.

10. If an organism has only three pairs of chromosomes, how many pairs of chromosomes are autosomes?
    a. one
    b. two
    c. three
    d. none

11. Labrador retrievers may have yellow fur, chocolate-brown fur, or black fur. Coat color is determined by two genes that have two alleles each. One gene, called E, determines whether the dog's fur is dark or light, with the dark allele (*E*) dominant to the light allele (*e*). The other gene, called B, determines whether the dark fur will be black or chocolate, with the black allele (*B*) dominant to the chocolate allele (*b*). If you mate two chocolate labs together, what color puppies can they have?
    a. Only chocolate puppies can be produced because both parents must be *EEbb*.
    b. Chocolate or yellow puppies can be produced because the parents can be either *Eebb* or *EEbb*.
    c. Chocolate, yellow, and black puppies will be produced because both parents must be *EeBb*.
    d. Chocolate and black puppies will be produced because both parents must be *EeBb*.

12. A couple brings home their new, nonidentical twin daughters, Joan and Jill. After several months, the father begins to suspect that there was a mix-up at the hospital, because Jill does not look much like either parent or like her sister. When the twins' blood tests come back, the father calls his lawyer to start a lawsuit against the hospital. The mother, father, and Joan have type A blood, but Jill has type O blood. Based on bloodtype, does the father have a case? Explain your answer. (The gene for blood type has three alleles: *A*, *B*, and *O*. The *A* and *B* alleles are codominant, and the *O* allele is recessive.)
    a. No, because parents with type A blood can have a child with type O blood.
    b. No, because parents with any blood type (A, B, AB, or O) can produce children with type O blood.
    c. Yes, because all of this couple's children will have type A blood.
    d. Yes, because people with type A blood can pass on only *A* alleles to their children.

13. An individual who is a "carrier" of a genetic disorder is
    a. heterozygous for the disorder, and the allele for the disorder is recessive.
    b. heterozygous for the disorder, and the allele for the disorder is dominant.
    c. homozygous for the disorder.
    d. homozygous for the disorder but is protected by the presence of an immunity gene.

14. A man with hemophilia inherited the hemophilia gene from
    a. his mother.
    b. his father.
    c. either his mother or his father.

15. Which of the following is caused by an abnormal number of autosomes?
    a. Down's syndrome
    b. Klinefelter syndrome
    c. Turner syndrome
    d. Marfan syndrome

ANSWER KEY

Quiz 1

1. b
2. c
3. c
4. b
5. d
6. e
7. a
8. e

9. b
10. a
11. b
12. d
13. c
14. d
15. b

Quiz 2

1. c
2. b
3. e
4. e
5. b
6. c
7. b
8. c

9. d
10. b
11. b
12. a
13. a
14. a
15. a

CHAPTER 13: BIOTECHNOLOGY

OUTLINE

Section 13.1 What Is Biotechnology?

- **Biotechnology** is any use or alteration of organisms, cells, or biological molecules to achieve specific practical goals. Modern biotechnology uses **genetic engineering** as a direct method to modify genetic material.
- Some genetic engineers transfer **recombinant DNA** (which can be grown in bacteria, viruses, or yeasts) to other organisms so that the trait carried by the DNA can be expressed. Organisms that express DNA derived from other species are called **genetically modified** (or **transgenic**) **organisms** (GMOs).

Section 13.2 How Does DNA Recombine in Nature?

- Recombinant DNA occurs naturally by three processes: (1) **Sexual reproduction** by crossing over, (2) bacterial **transformation** by the acquisition of DNA from **plasmids** or other bacteria (**Figure 13-1**), or (3) viral infection, when DNA fragments are transferred between or among species (**Figure 13-2**).

Section 13.3 How Is Biotechnology Used in Forensic Science?

- Small amounts of DNA can be amplified into large amounts by the **polymerase chain reaction** (**PCR**) technique (**Figure 13-3**). This DNA can be cut into specific fragments, such as **specific tandem repeats** (STRs), using restriction enzymes (**Figure 13-4**).
- DNA fragments (such as STRs) can be separated by gel electrophoresis (**Figure 13-5**), which are then made visible by the use of **DNA probes** (**Figure 13-6**). STRs patterns are unique for each individual, creating a **DNA profile**, which can be used to match crime scene DNA to that of a suspect.

Section 13.4 How Is Biotechnology Used in Agriculture?

- Crop plants have been modified for herbicide or pest resistance. This can be accomplished by (1) using restriction enzymes to insert the desired gene into a plasmid, (2) transforming bacteria that are then used to infect plant cells, and (3) culturing them to form whole transgenic plants (**Figures 13-9** and **13-10**). Transgenic plants can also be used to produce a variety of different medicines.
- Transgenic animals can also be produced (often using a disabled virus to transfer DNA) with enhanced growth, milk production, or the ability to produce human proteins.

Section 13.5 How Is Biotechnology Used to Learn About the Human Genome?

- Biotechnology techniques were used to sequence the human genome. This knowledge is used to discover medically important genes, to examine evolutionary relatedness among organisms, and to study genetic variability among individuals.

Section 13.6 How Is Biotechnology Used for Medical Diagnosis and Treatment?

- Restriction enzymes are used to produce DNA segments at a variety of lengths (called **restriction fragment length polymorphisms—RFLPs**) to identify alleles for a specific genetic disorder (e.g., sickle-cell anemia, **Figure 13-11**).
- DNA probes can also be used to identify alleles for specific genetic disorders (e.g., cystic fibrosis, **Figure 13-12**).
- Genetic engineering can be used to insert functional alleles into normal cells, stem cells, or egg cells to correct the genetic disorder.

Section 13.7 What Are the Major Ethical Issues of Modern Biotechnology?

- One controversial aspect of Genetically Modified Organisms (GMOs) is consumer safety. GMOs usually contain proteins that are harmless to mammals, but there is the potential that new allergens can be transferred to food that is normally harmless. This can be avoided by thorough testing prior to commercial distribution.
- A second controversial aspect of GMOs is environmental protection. Modified plant genes can be transferred to wild plants, possibly disrupting ecosystems and agriculture. Also, mobile transgenic animals could displace wild populations.
- Use of biotechnology on human embryos is controversial (**Figure 13-13**). The primary benefit is that genetic disorders can be cured, but it can also mean that embryos with genetic defects can be detected and the pregnancies terminated. Potentially, the genomes of human embryos can be modified to enhance their traits rather than to fix any serious defects.

LEARNING OBJECTIVES

Section 13.1 What Is Biotechnology?

- Explain the relationship between recombinant DNA and transgenic organisms.

Section 13.2 How Does DNA Recombine in Nature?

- Discuss the different examples of natural genetic recombination.

Section 13.3 How Is Biotechnology Used in Forensic Science?

- Describe the steps in polymerase chain reaction (PCR).
- Explain how short tandem repeats are used as a genetic fingerprint of an individual.
- Explain how DNA fragment size is determined by gel electrophoresis.

Section 13.4 How Is Biotechnology Used in Agriculture?

- Explain the steps involved in creating a genetically modified plant resistant to insects.
- Explain how restriction enzymes are used to move a gene from one organism to another.
- Discuss the possible future roles of genetically modified plants and animals in medicine.

Section 13.5 How Is Biotechnology Used to Learn About the Human Genome?

- Discuss the various benefits of the Human Genome Project.

Section 13.6 How Is Biotechnology Used for Medical Diagnosis and Treatment?

- Explain how restriction fragment length polymorphisms (RFLP) analysis is used to determine the sickle genotype of an individual.
- Describe how complementary base pairing allows for diagnosis of cystic fibrosis.
- Discuss how biotechnology has enhanced the treatment of diseases.
- Explain the difference in outcome between inserting a corrected gene into an adult cell and into a stem cell.

Section 13.7 What Are the Major Ethical Issues of Modern Biotechnology?

- Discuss the ethics of altering an organism's genome.
- Discuss the possible ramifications of genetically altered organisms "escaping" into the wild-type population.
- Discuss the possible environmental effects of transgenic organisms.

QUIZ 1

1. Genetic engineering
   a. is more than 10,000 years old.
   b. can be used only on bacteria.
   c. is designed to randomly mix chromosomes from different organisms.
   d. can be implemented to improve livestock for agricultural uses.

2. Piecing together genes from different organisms is referred to as
   a. transmorphic DNA.
   b. transgenic DNA
   c. recombinant DNA.

3. DNA recombination occurs
   a. during sexual reproduction.
   b. only naturally.
   c. only in the laboratory.
   d. with vectors called plasmids.
   e. only in bacteria.

4. Transformation is the process by which foreign _____ is taken up from a cell's environment, permanently changing the characteristics of a cell and its offspring.
   a. DNA
   b. RNA
   c. protein

5. Antibiotic resistance can be transferred from one bacterial strain to another by _____.
   a. transformation
   b. natural selection
   c. transplantation

6. Which of the following describes an example of DNA recombination?
   a. any organism that utilizes sexual reproduction
   b. the exchange of fluids between a cell's interior (cytoplasm) and its exterior (extracellular fluid)
   c. a viral infection resulting in a common cold
   d. (a) and (c)

7. What is the polymerase chain reaction (PCR) technique?
   a. a technique that incorporates repeated cycles of heating and cooling of DNA segments in the presence of primers and heat-resistant DNA polymerase
   b. an elegant chemical technique that utilizes the natural ability of DNA polymerase to synthesize DNA that is complementary to the parent DNA fragments
   c. a technique that allows investigators to determine the nucleotide sequence of a given gene
   d. a technique that has led to the development of genetic screening tests of newborns for genetic disease (e.g., cystic fibrosis) once the nucleotide sequence of disease-causing gene is known
   e. all of the above

8. Gel electrophoresis is used to _____.
   a. increase the amount of DNA of a particular sequence
   b. cleave DNA into small pieces
   c. separate DNA fragments by size
   d. extract DNA from tissue samples

9. Researchers must design short pieces of DNA that are complementary to DNA on either side of the segment to be amplified. These small pieces are called _____.
   a. primers
   b. DNA probes
   c. plasmids
   d. DNA fingerprints

10. Which of the following sources of DNA could be used in DNA forensic analysis?
    a. semen
    b. saliva
    c. hair follicle
    d. blood
    e. all of the above

11. What are restriction enzymes?
    a. enzymes that are limited in how big they can become
    b. vitamins, such as vitamin A
    c. plasmids
    d. enzymes that cleave through a DNA helix wherever they encounter a specific sequence of nucleotides

12. The bulk of genetically modified organisms are _____.
    a. animals
    b. plants
    c. bacteria
    d. fungi

13. The primary goal of human gene therapy is to _____.
    a. correct genetic disorders by inserting normal genes in place of defective ones
    b. provide counseling to affected people to help them live with their disorders
    c. treat the symptoms of genetic disorders
    d. correct genetic disorders in developing embryos

14. Why is stem cell technology a promising area of medical research?
    a. Stem cell DNA is easy to fingerprint.
    b. Under the right conditions, a stem cell could become any type of cell in the body.
    c. Stem cells can fight pathogens.
    d. Stem cells can be used to produce new medicines.

15. How might human cloning permanently correct genetic defects?

    a. A viral vector would replace a defective gene and the resulting corrected nucleus would be injected into an egg lacking a nucleus and then allowed to develop *in utero*.

    b. Cloning many identical humans would increase the chance that mutation would correct the genetic defect.

    c. Replacement humans could be grown to be used when the original succumbs to a genetic disease.

    d. All of the above apply.

QUIZ 2

1. Transgenic organisms can
    a. contain only DNA from their same species.
    b. express DNA from another species.
    c. only be primates.
    d. only be bacteria.
    e. express only their own DNA.

2. All of the following are true of plasmids EXCEPT that they
    a. are very small compared to the bacterial chromosome.
    b. are found in bacteria, protists, fungi, and algae.
    c. can contain genes providing their host bacteria with resistance to antibiotics.
    d. can be present in many more copies per cell than the chromosome.
    e. do not carry any genes.

3. Which process involves the movement of plasmids from one bacteria into another?
    a. crossing over
    b. viral infection
    c. recombinant DNA library formation
    d. transformation

4. Transformation of bacteria occurs when bacteria take up new
    a. viruses.
    b. only DNA fragments.
    c. only plasmids.
    d. DNA fragments or plasmids

5. In bacterial transformation, after a plasmid enters the new bacterium, it
    a. is destroyed.
    b. will replicate in the cytoplasm.
    c. will be incorporated into the bacterial chromosome.
    d. causes the bacterium to die.
    e. will be broken into small DNA pieces.

6. The polymerase chain reaction
    a. is a method of synthesizing human protein from human DNA.
    b. takes place naturally in bacteria.
    c. can produce billions of copies of a DNA fragment in several hours.
    d. uses restriction enzymes.
    e. is relatively slow and expensive compared with other types of DNA purification.

7. Imagine you are looking at a DNA fingerprint that shows an STR pattern of a mother's DNA and her child's DNA. Will all the bands on the Southern Blot showing the child's DNA match those of the mother? Explain. (To help you answer this question, visit the Biology Animation Library at Cold Spring Harbor Laboratory Web site and play the tutorial about the Southern Blot technique.)
    a. Yes, because the child developed from an egg produced by the mother.
    b. Yes, because the DNA of mothers and children is identical.
    c. No, because a person's DNA pattern changes with age.
    d. No, because the father contributed half the child's DNA.

8. STR polymorphisms can be used to identify individuals because _____.
    a. individuals are genetically unique
    b. restriction enzymes cut different recognition sequences in different people
    c. a given set of PCR primers works for only a small subset of individuals

9. DNA fingerprinting
    a. requires large amounts of DNA.
    b. is useful only for forensic analysis.
    c. can involve analysis of RFLPs.
    d. can only prove guilt, never innocence.
    e. is a constitutional right.

10. Why are genes inserted into bacterial plasmids during the gene-cloning process?
    a. Large amounts of the gene can be produced when the bacteria multiply.
    b. Genes can be cloned only for bacteria.
    c. The genes are more stable when inserted into bacterial plasmids.
    d. Bacterial plasmids require inserted genes to be functional.

11. Why might it be beneficial to produce genetically modified animals?
    a. We can populate our zoos with bizarre and interesting animals.
    b. We may be able to develop livestock with beneficial characteristics.

    c. We may be able to develop animals that can produce useful medicines.
    d. (b) and (c) are correct.

12. Analysis of restriction fragment length polymorphisms (RFLPs) is a rapid way to examine differences in the _____.
    a. bacterial enzymes
    b. length of a DNA molecule
    c. DNA sequences of individuals
    d. number of genes on a chromosome

13. What benefit does administration of genetically engineered human insulin provide that pig/bovine insulin does not?
    a. There was no benefit; the nucleotide sequence that determines the structure of insulin is the same in all three species.
    b. Allergic reactions were avoided by using the recombinant human insulin.
    c. It could be produced in large quantities using bacteria/yeast genetically engineered to carry the human insulin gene.
    d. (b) and (c) are correct.

14. What are some of the problems associated with the use of genetic screens for cystic fibrosis?
    a. Positive testing of a man and a woman each carrying a copy of the CF gene guarantees their offspring will have the disease.
    b. Ethical dilemmas: Should individual carriers of the CF allele attempt to conceive? Should society and insurance carriers be forced to bear the financial costs?
    c. There is the possibility of genetic discrimination by insurance companies.
    d. (B) and (c) are correct.

15. What ethical argument(s) has(have) been used against biotechnology?
    a. Genetically modified foods are dangerous to eat.
    b. Genetically modified organisms are harmful to the environment.
    c. The human genome should not be changed because the repercussions of doing so are unclear.
    d. All of the above.

ANSWER KEY

Quiz 1

1. d		**9.** a	
2. c		**10.** e	
3. a		**11.** d	
4. a		**12.** b	
5. a		**13.** a	
6. d		**14.** b	
7. e		**15.** a	
8. c			

Quiz 2

1. b		**9.** c	
2. e		**10.** a	
3. d		**11.** d	
4. d		**12.** c	
5. b		**13.** d	
6. c		**14.** d	
7. d		**15.** d	
8. a			

CHAPTER 14: PRINCIPLES OF EVOLUTION

OUTLINE

Section 14.1 How Did Evolutionary Thought Evolve?

- Pre-Darwinian science stated that God simultaneously created all organisms, each being distinct and unchanging from the moment of its creation. This view was reflected in Aristotle's "Ladder of Nature" (**Figure 14-2**).
- By the 1700s, naturalists had explored many lands and discovered more species than they expected. They also noticed that physical and geographic comparisons of these species seemed inconsistent with the supposed "fixed" nature of life.
- LeClerc suggested that God created a relatively small number of species and that modern species were "produced by Time,"—they evolved through natural processes.
- The discovery and examination of **fossils** suggested that previously unknown organisms had gone extinct, and that fossil remains showed a remarkable progression over time (**Figures 14-3** and **14-4**).
- The theory of **catastrophism** (proposed by Cuvier) suggested that these unknown species were eliminated in successive catastrophes that deposited their fossil remains in rock layers. **Uniformitarianism** (proposed by Lyell) contradicted this view, suggesting that natural, observable processes, repeated over long periods of time, could also produce the layers of rock in which fossils were found. This also suggested that Earth was many millions of years old.
- Lamarck suggested that organisms evolved through the **inheritance of acquired characteristics**, in which organisms modify parts through use or disuse and pass them on to offspring. In contrast, Darwin and Wallace proposed that organisms evolve over time through **natural selection**.

Section 14.2 How Do We Know That Evolution Has Occurred?

- The fossil record shows evidence of evolutionary change over time (**Figure 14-6**).
- **Homologous structures**, **vestigial structures**, and **analogous structures** all provide evidence of descent with modification (**Figures 14-7**, **14-8**, and **14-9**).
- Similarities among vertebrate embryos suggest common ancestry (**Figure 14-10**).
- Genetic and biochemical analyses suggest that all organisms are related to different degrees (**Figure 14-11**).

Section 14.3 How Does Natural Selection Work?

- Darwin and Wallace suggest that (1) populations are variable, (2) variable traits can be inherited, (3) some individuals in a population survive and others do not, and then (4) the characteristics of successful individuals will be "naturally selected" and become more common over time.

Section 14.4 What Is the Evidence That Populations Evolve by Natural Selection?

- Artificial selection supports evolution by natural selection (**Figure 14-13**). Since humans have bred different forms of plants and animals over thousands of years, logic suggests that natural selection can produce the entire spectrum of organisms over hundreds of millions of years.
- Natural selection occurs today, as observed in some guppy populations (**Figure 14-14**), insect pesticide resistance, and scientific experimentation.

LEARNING OBJECTIVES

Section 14.1 How Did Evolutionary Thought Evolve?

- Discuss the early theories on the diversity of life on the planet.
- Describe the connections among fossils, rock layers, and geological time.
- Explain the theory of catastrophism.
- Explain the flaw in Lamarck's theory of acquired characteristics.

Section 14.2 How Do We Know That Evolution Has Occurred?

- Discuss how fossil evidence and comparative anatomy support the theory of natural selection.
- Discuss how DNA sequencing supports evolution.

Section 14.3 How Does Natural Selection Work?

- Explain the postulates of natural selection.

Section 14.4 What Is the Evidence That Populations Evolve by Natural Selection?

- Explain the difference between artificial and natural selection.
- Discuss the role of the environment on natural selection.

QUIZ 1

1. The idea put forward by Lamarck, that organisms could pass on to their offspring physical changes that the parents developed during their own lifetimes, is known as _____.
   a. genetic drift
   b. natural selection
   c. artificial selection
   d. adaptive radiation
   e. inheritance of acquired characteristics

2. Which of the following is a main point of Darwin's theory of evolution?
   a. Life on Earth is quite old.
   b. Evolution is gradual and continuous.
   c. Contemporary species share a common descent.
   d. Species are formed and adapt by the process of natural selection.
   e. All of the above are correct.

3. What does the idea of *uniformitarianism* suggest about the geological record and the age of Earth?
   a. Earth is 6000 years old.
   b. Species are evidence of acts of divine creation.
   c. Earth's species were created initially but many were destroyed by successive catastrophes.
   d. Natural processes (e.g., sedimentation due to river flow) occurring over long stretches of time rather than cataclysmic events account for the thick layers of rock where fossils are found.

4. What was the untested weakness in Darwin's *On the Origin of Species*?
   a. Organisms evolved through the inheritance of acquired characteristics.
   b. Organisms evolved through natural selection.
   c. Traits passed from generation to generation, and variations in these traits occurred randomly.
   d. All of the above are correct.

5. A human arm is homologous with a (an)
   a. seal flipper.
   b. octopus tentacle.
   c. bird wing.
   d. sea star arm.
   e. (a) and (c)

6. Which of the following are fossils?
   a. pollen grains buried in the bottom of a peat bog
   b. the petrified cast of a clam's burrow
   c. the impression a clam shell made in mud, preserved in mudstone
   d. an insect leg sealed in plant resin
   e. all of the above

7. In Africa, a species of bird called the yellow-throated longclaw looks almost exactly like the meadowlark found in North America, but they are not closely related. This is an example of
   a. uniformitarianism.
   b. gradualism.
   c. vestigial structures.
   d. convergent evolution.

8. Bat wings and insect wings are _____ structures.
   a. analogous
   b. homologous

9. All organisms share the same genetic code. This commonality is evidence that
   a. evolution is occurring now.
   b. convergent evolution has occurred.
   c. evolution occurs gradually.
   d. all organisms descended from a common ancestor.
   e. life began a long time ago.

10. Which of the following is a basic requirement for natural selection to be an effective evolutionary force?
    a. Mutation must occur frequently.
    b. Individuals reproduce at a rapid rate.
    c. Each population is limited to a small size.
    d. A population exhibits some genetic variability.
    e. All of the above are correct.

11. Which of the following is an INCORRECT statement about mutation?
    a. Mutation introduces variation into a population.
    b. Mutations can be inherited from parents to offspring.
    c. Mutations may have no effect on the organism.
    d. Mutations that are favored by selection are more likely to occur.

12. For cockroaches to have rapidly evolved resistance to the insecticide Combat®
    a. they must have learned to avoid it.
    b. some cockroaches must have had a mutation, causing them to dislike the ingredients.
    c. few people must have been using this brand.

13. Antibiotic prescriptions normally specify that one take the entire course of treatment, such as a fixed number of pills, rather than simply taking the medicine until one feels better. Why?
    a. It is wasteful to leave pills behind.
    b. You may suffer a relapse of the illness.
    c. Bacteria carrying slight resistance will be killed by the full course but will persist with a lesser dose.

14. Biologists have shown that some species of fish have genetic variations in their tendency to bite an angler's bait and hook. In professional bass-fishing tournaments, all fish caught are returned to the water; therefore
    a. catching fish should become more difficult.
    b. catching fish should become easier.
    c. no change in catching success should occur.

15. Behavioral geneticists have found that the ability of house mice to squeak is determined by a single gene. If squeaky mice are more likely to be dropped by startled predators
    a. the mouse population will consist of more and more squeakers over time.
    b. the mouse population will consist of more and more nonsqueakers over time.
    c. the occurrence of squeaking in the mouse population will not change.
    d. squeaking will become louder and more frequent over time.

QUIZ 2

1. Which of the following people did NOT influence Charles Darwin in his formulation of natural selection?
   a. Thomas Malthus
   b. Charles Lyell
   c. William Smith
   d. Gregor Mendel

2. Darwin contributed to science by recognizing and describing
   a. the process of natural selection.
   b. the pattern of evolution.
   c. many new species in South America and on Pacific islands.
   d. all of the above.

3. What is a modern definition of *evolution*?
   a. the formation of a new species
   b. a change in the characteristics of a population over time
   c. natural selection for a novel trait
   d. a trait that increases the fitness of an individual relative to individuals without that trait.

4. Which of the following lines of evidence support(s) the idea that evolution occurs?
   a. the fossil record
   b. genetic and biochemical analyses
   c. comparative anatomy and embryology
   d. artificial selection
   e. all of the above

5. Structures that have similar functions and superficially similar appearance but very different anatomy, such as the wings of insects and birds, are called _____.
   a. analogous structures
   b. homologous structures
   c. vestigial structures

6. Structures that may differ in function but have similar anatomy, presumably because of descent from common ancestors, are called _____.
   a. analogous structures
   b. homologous structures
   c. vestigial structures

7. Structures that serve no apparent purpose but are homologous to functional structures in related organisms are called _____.
   a. analogous structures
   b. homologous structures
   c. vestigial structures

8. All vertebrate embryos resemble each other during the early stages of development. For example, fish, turtles, chickens, mice, and humans develop tails and gill slits during early stages of development. This suggests that _____.
   a. early embryonic development is conservative
   b. ancestral vertebrates possessed genes that directed the development of tails and gill slits, and all of their descendants still retain those genes

c. genes that modify the developmental pathways in vertebrates arose later in evolution

d. all of the above

9. Which of the following is an example of analogous structures?

a. molar teeth in a vampire bat

b. superficially similar structures in unrelated species

c. internally similar structures (e.g., of birds and mammals) that are used for many different functions

d. (a) and (c)

10. You are a biologist studying a natural population of mice, and you observe that in one area the proportion of darker-colored mice is greater than the proportion of lighter-colored mice. In another area, the opposite is true. You find that only the area with more dark mice has predators. Therefore, you hypothesize that darker mice are favored in areas with predators, perhaps because they are more difficult to see. If your hypothesis is true, what would you expect to happen in the few generations after predators are introduced to an area with a population of mice that previously did not have predators?

a. The proportion of darker-colored mice will decrease.

b. The proportion of darker-colored mice will not change.

c. The proportion of darker-colored mice will increase.

d. The predators will evolve to be able to see the darker mice better.

11. Which list contains the four observations of natural selection?

a. variation in population, heritable variation, selection, unlimited survival and reproduction

b. uniform population, heritable traits, selection, differential survival or reproduction

c. variation in population, environmental variation, selection, differential survival or reproduction

d. rapid reproduction rate, constant population size, variation among individuals in a population, heritable variation

12. Which of the following would describe artificial selection?

a. breeding organisms for the purpose of generating certain features or traits

b. coloration changes in guppy populations in the absence of predators

c. increased frequency of roaches that avoid sugar-baited poison traps

d. all of the above

13. In what way does the human population influence evolution?

a. Human development changes the habitats of many species, influencing natural selection on those species.

b. Use of antibiotics by humans has selected for anitibiotic-resistant bacterial populations.

c. Humans are responsible for the many breeds of dogs found today.

d. All of the above are correct.

14. In the United States today, about half the corn crop is genetically engineered with a protein that is toxic to corn borers, an insect pest of corn. Which of the following conditions is necessary for the corn borer to evolve resistance to the toxic protein?

a. The corn borer must have heritable variation in the resistance to the toxic protein. The resistant corn borers must survive better or reproduce more than nonresistant corn borers.

b. The corn borers must lack variation in the resistance to the toxic protein. The resistant corn borers must survive in the same way as nonresistant corn borers.

c. All corn borers must have resistance to the toxic protein. The resistant corn borers must survive better or reproduce more than other insects.

d. The corn borer must have heritable variation in the resistance to the toxic protein. The resistant corn borers must survive and reproduce the same as nonresistant corn borers.

15. Biologists have shown that some species of fish have genetic variation in their tendency to bite an angler's bait and hook. With repeated fishing and harvesting of caught fish

a. catching fish should become more difficult.

b. catching fish should become easier.

c. no change in catching success should occur.

d. an individual's ability to catch fish relative to other anglers will determine one's success.

ANSWER KEY

Quiz 1

1. e
2. e
3. d
4. c
5. e
6. e
7. d
8. a

9. d
10. d
11. d
12. b
13. c
14. c
15. a

Quiz 2

1. d
2. d
3. b
4. e
5. a
6. b
7. c
8. d

9. b
10. d
11. d
12. a
13. d
14. a
15. a

CHAPTER 15: HOW ORGANISMS EVOLVE

OUTLINE

Section 15.1 How Are Populations, Genes, and Evolution Related?

- Evolution is described as a change in a population's **allele frequencies** over time, resulting in a change in that population's **gene pool** (**Figure 15-2**).
- Allele frequencies in a population will remain stable over successive generations (forming an **equilibrium population**) only if the following conditions are met: (1) There are no mutations. (2) There is no gene flow between populations. (3) The population must be very large. (4) All mating must be random. (5) There must be no natural selection. These conditions are rarely met.

Section 15.2 What Causes Evolution?

- **Mutations** are the original source of genetic variability and are usually the result of DNA copying errors during cell division. These mutations are random events and may be passed on to successive generations (**Figure 15-3**).
- **Gene flow** describes the movement of alleles between populations. This can change population allele frequencies by increasing the genetic similarity of these populations.
- Certain events have the potential to cause random changes in a population's allele frequency by a process called **genetic drift** (**Figure 15-5**). Small populations are particularly affected because these events can eliminate a disproportionate number of individuals bearing a particular allele (**Figure 15-6**). Both population bottlenecks and the founder effect are examples of situations in which genetic drift can occur (**Figures 15-7** and **15-8**).
- Nonrandom mating can affect the genotypes of a population, typically by **inbreeding** (which can result in harmful homozygous genotypes) or **assortative mating**.
- Not all genotypes are equally beneficial. Different genotypes result in different phenotypes, which affect the ability of an organism to survive in different ways. Thus, natural selection will favor the genotypes that produce beneficial phenotypes.

Section 15.3 How Does Natural Selection Work?

- Natural selection acts upon a population through differences in reproductive success resulting from phenotype variations. This, in turn, affects a population's genotype.
- Natural selection occurs due to interactions between organisms and the **abiotic** or **biotic** components of the environment. An example of abiotic influences on natural selection is a plant's ability to access soil minerals and soil water content. Biotic influences on natural selection include **competition** for resources, **coevolution** due to the interactions between predators and prey, and **sexual selection** (**Figures 15-11** and **15-12**).
- Both natural and sexual selection can result in three patterns of evolutionary change (**Figure 15-13**).
  (1) **Directional selection** shifts character traits in a specific direction. (2) **Stabilizing selection** shifts character traits toward the average for a population. (3) **Disruptive selection** shifts character traits toward the phenotypic extremes for a population.

LEARNING OBJECTIVES

Section 15.1 How Are Populations, Genes, and Evolution Related?

- Explain the relationship between allele frequencies and a population's gene pool.
- Explain the Hardy-Weinberg principle.

Section 15.2 What Causes Evolution?

- Describe the factors that change the allele frequency of a population.
- Explain how gene flow affects allele frequencies.
- Explain the role of population size on maintaining allele frequencies.
- Describe the difference between a population bottleneck and the founder effect.

- Explain why nonrandom mating increases the number of homozygous individuals.
- Explain how natural selection accounts for the emergence of antibiotic-resistant organisms.

Section 15.3 How Does Natural Selection Work?

- Explain the role of reproduction in natural selection.
- Discuss the role of adaptation on reproduction and natural selection.
- Discuss the various factors influencing selection within a community of different populations.
- Explain the difference between directional, stabilizing, and disruptive selection.
- Explain the persistence of the sickle cell allele in equatorial regions.

QUIZ 1

1. What is a gene pool?
   a. a region of DNA found at a specific position on a chromosome
   b. the number of copies of an allele for a specific gene in a population
   c. the total number of all the genes in a population
   d. none of the above

2. The Hardy-Weinberg principle is a model of population genetics that
   a. shows that evolution will occur if all the conditions of the principle are met.
   b. shows that evolution will occur if even one of the conditions of the principle is violated.
   c. describes how evolution occurs in nature.
   d. (a) and (c) are correct.

3. Which of the five conditions of the Hardy-Weinberg principle describes how unique, noninherited DNA can enter a population?
   a. mutations
   b. decreasing population size
   c. nonrandom mating
   d. natural selection
   e. no gene flow

4. Evolution is best defined as a change in _____.
   a. number of species
   b. physical traits
   c. DNA sequence
   d. allele frequencies

5. A large population of birds on an island has a color distribution of 80% green birds and 20% yellow birds. Most of the population is killed by an unusually strong storm; color distribution of the survivors is 60% green birds and 40% yellow birds. The change in the color distribution is probably due to
   a. natural selection.
   b. gene flow.
   c. genetic drift.
   d. mutation.

6. Mutations
   a. are responsible for rapid evolution in populations.
   b. are defined as changes in single DNA nucleotides.

   c. provide the basic "raw material" for evolution.
   d. are usually beneficial to the organism.

7. Gene flow _____.
   a. cannot influence the evolution of a population
   b. prevents the spread of alleles through a species
   c. causes populations to diverge from each other
   d. makes populations more genetically similar

8. Which of the following is characterized by random changes in allele frequencies?
   a. natural selection
   b. genetic drift
   c. gene flow
   d. nonrandom mating

9. When alleles increase in frequency because they allow the organism to better survive and reproduce in its environment, this is a result of:
   a. natural selection.
   b. gene flow.
   c. genetic drift.
   d. mutation.

10. Selection against individuals at both ends of a phenotypic distribution for a character (i.e., favoring those in the middle or average of the distribution) is an example of _____.
    a. directional selection
    b. disruptive selection
    c. stabilizing selection

11. The process by which two species evolve adaptations in response to each other, such that an evolutionary change in one species produces an evolutionary change in the other, is called _____.
    a. coevolution
    b. natural selection
    c. directional selection
    d. balanced polymorphism

12. Which of the following selective processes did Darwin suggest to explain the evolution of the conspicuous structures and courtship behaviors that male animals use to attract a mate?
    a. kin selection
    b. sexual selection
    c. natural selection

13. Which of the following is most likely to result in the evolution of altruistic behavior?
    a. kin selection
    b. sexual selection
    c. natural selection

14. Black-bellied seedcrackers have either small beaks (better for eating soft seeds) or large beaks (better for hard seeds). There are no seeds of intermediate hardness; therefore, _____ acts on beak size in seedcrackers.
    a. directional selection
    b. stabilizing selection
    c. disruptive selection

15. Goldenrod gall flies lay their eggs in the goldenrod plant. The fly larvae cause the plant to form a gall, a swollen growth of plant tissue. Large galls tend to be pecked open and their larvae eaten by birds. Small galls are more frequently parasitized by tiny wasps, whose larvae eat the fly larva and take over the gall. This complex situation implies that _____ is acting on gall size.
    a. directional selection
    b. stabilizing selection
    c. disruptive selection

QUIZ 2

1. A _____ is all the individuals of a species living in a particular place at a particular time.
   a. bottleneck
   b. species
   c. gene pool
   d. population

2. The relative proportion of an allele in a population is the
   a. allele frequency.
   b. gene pool.
   c. Hardy-Weinberg equilibrium.
   d. population size.

3. An equilibrium population
   a. evolves slowly.
   b. evolves quickly.
   c. evolves only by natural selection.
   d. does not evolve.

4. Which of the following statements is correct?
   a. Natural selection is a mindless, mechanical process.
   b. Darwin had no knowledge of the mechanics of heredity.
   c. Heredity plays an important role in the theory of evolution by natural selection.
   d. Evolution is a property not of individuals but of populations.
   e. The changes that we see in an individual as it grows and develops are not evolutionary changes.
   f. All of the above are correct.

5. Which of the following is a mechanism or cause of evolution?
   a. mutation
   b. gene flow
   c. genetic drift
   d. natural selection
   e. all of the above

6. In a mainland bird population, most individuals are black in color, and gray is a rare variation. A small group of these birds is carried by strong winds to a distant island, where they establish a new population. After a few generations, the gray phenotype is very common. What is most likely to be responsible for this?
   a. natural selection
   b. mutation
   c. gene flow
   d. nonrandom mating
   e. founder effect

7. Natural selection acts (through predation) against banded water snakes on certain Lake Erie islands, favoring the uniformly light-colored snakes. The banded form is very common on the nearby mainlands. Yet banded snakes are maintained in the island populations and not eliminated completely. This is probably due to _____.
   a. mutation
   b. gene flow
   c. genetic drift
   d. natural selection

8. Genetic drift will tend to _____.
   a. increase genetic variability both within and between populations
   b. decrease genetic variability both within and between populations
   c. increase genetic variability within populations but decrease genetic variability between populations
   d. decrease genetic variability within populations but increase genetic variability between populations

9. Which of the following does NOT change allele frequencies in a population?
   a. mutation
   b. gene flow

c. genetic drift
d. random mating
e. natural selection

10. The Hardy-Weinberg equilibrium represents an idealized, evolution-free population in which the allele frequencies and genotype frequencies will not change over time. In order for this to happen, five conditions must be met: (1) there must be no mutation; (2) there must be no gene flow between populations; (3) the populations must be very large; (4) all mating must be random; and (5) there must be no natural selection. If one of these five conditions were violated, genetic change, and thus evolution, would occur in the populations of subsequent generations. Suppose that only condition 3 were violated—that the population was very small. In this situation, the evolution would probably be due to _____.
    a. mutation
    b. migration
    c. genetic drift
    d. natural selection

11. Which phrase best describes the concept of natural selection?
    a. survival of the fittest
    b. reproductive success
    c. long life
    d. (a) and (b)

12. Which type of natural selection favors individuals with traits that vary from the norm over individuals with traits that are frequently encountered?
    a. disruptive selection
    b. stabilizing selection
    c. directional selection

13. Body size varies among individuals in a species of lizard in the genus *Aristelliger*. Small lizards have a hard time defending a territory, and thus mating, but large lizards are more likely to be preyed on by owls. Therefore, natural selection favors individuals with an average body size. This is an example of _____.
    a. directional selection
    b. disruptive selection
    c. stabilizing selection

14. If a population is undergoing natural selection, then
    a. its gene frequencies are not changing.
    b. it is not becoming more adapted to its environment.
    c. it is becoming more adapted to its environment.

15. Long necks make it easier for giraffes to reach leaves high on trees, while also making them better fighters in "neck-wrestling" contests. In both cases, _____ appears to have made giraffes the long-necked creatures they are today.
    a. directional selection
    b. stabilizing selection
    c. disruptive selection

ANSWER KEY

Quiz 1

1. c	**9.** a
2. b	**10.** c
3. a	**11.** a
4. d	**12.** b
5. c	**13.** a
6. c	**14.** c
7. d	**15.** b
8. b	

Quiz 2

1. d
2. a
3. d
4. f
5. e
6. e
7. b
8. d
9. d
10. c
11. b
12. a
13. c
14. c
15. a

CHAPTER 16: THE ORIGIN OF SPECIES

OUTLINE

Section 16.1 What Is a Species?

- The **biological species concept** defines a species as all the groups of actually or potentially interbreeding natural populations that are also **reproductively isolated** from other populations.

Section 16.2 How Is Reproduction Isolation Between Species Maintained?

- **Isolating mechanisms** prevent species interbreeding by maintaining reproductive isolation (**Table 16-1**).
- **Premating isolation mechanisms** prevent mating between species. These mechanisms include **geographical isolation (Figure 16-3)**, **ecological isolation (Figure 16-4)**, **temporal isolation (Figure 16-5)**, **behavioral isolation (Figure 16-6)**, and **mechanical incompatibility (Figure 16-7)**.
- **Postmating isolation mechanisms** prevent the formation of vigorous, fertile hybrids between species. These mechanisms include **gametic incompatibility**, **hybrid inviability**, and **hybrid infertility (Figure 16-8)**.

Section 16.3 How Do New Species Form?

- **Speciation** occurs when gene flow between two populations is reduced or eliminated and the populations diverge genetically.
- Speciation can occur through geographical separation, a process called **allopatric speciation (Figure 16-9)**.
- Speciation can also occur through ecological isolation, a process called **sympatric speciation (Figure 16-10)**.
- In some cases, many new species can form in a relatively short period of time, such as when a species encounters a wide variety of unoccupied habitats. This is called **adaptive radiation (Figure 16-13)**.

Section 16.4 What Causes Extinction?

- **Extinction** occurs when all the members of a species die off. This can occur from localized distribution (**Figure 16-14**) and overspecialization, competition among species, and habitat destruction.

LEARNING OBJECTIVES

Section 16.1 What Is a Species?

- Explain the term *reproductive isolation* and how it relates to the definition of a species.
- Discuss the characteristics commonly used to distinguish different species.

Section 16.2 How Is Reproduction Isolation Between Species Maintained?

- Describe isolating mechanisms and how they maintain reproductive isolation.
- Explain how postmating isolating mechanisms maintain reproductive isolation.
- Discuss the various premating isolating mechanisms.
- Describe the difference between geographic and habitat isolation mechanisms.
- Discuss the possible outcomes for hybrid offspring.

Section 16.3 How Do New Species Form?

- Explain the two factors required for speciation.
- Describe the different pathways leading to allopatric speciation.
- Discuss the role of natural selection in speciation.
- Describe the difference between sympatric speciation and allopatric speciation.
- Describe the conditions necessary for adaptive radiation.

Section 16.4 What Causes Extinction?

- Describe the environmental factors that drive species to extinction.
- Describe the roles species distribution and specialization play in extinction.
- Explain how competition and predation lead to extinction.
- Discuss the current effect human development has on extinction rates.

QUIZ 1

1. Which of the following is true of the biological-species concept?
   a. Asexually reproducing organisms use the same criteria as sexually reproducing organisms.
   b. Naturally occurring populations must actually interbreed to be considered as the same species.
   c. Different appearance is sufficient justification to categorize overlapping, naturally occurring populations as different species.
   d. None of the above is correct.

2. Biologists sometimes combine two groups previously considered different species into the same species because the two species have evolved to _____.
   a. be physically divergent, but they are found to interbreed in nature.
   b. lose premating reproductive-isolating mechanisms.
   c. lose postmating reproductive-isolating mechanisms.
   d. be more physically similar.
   e. expand their ranges and now encounter each other.

3. Why is appearance a poor characteristic to use when determining if two organisms are of the same or different species?
   a. Organisms of differing appearance may be different forms of the same species that can interbreed.
   b. Organisms of similar appearance may have mating incompatibilities.
   c. Ugly organisms are unlikely to mate with anything and therefore is a reproductive barrier.
   d. (a) and (b) are correct.

4. Incompatibilities that prevent mating between species are called
   a. premating isolation mechanisms.
   b. postmating isolation mechanisms.

5. In many species of fireflies, males flash to attract females. Each species has a different flashing pattern. This is probably an example of
   a. premating isolation.
   b. postmating isolation.

6. A female fig wasp will carry fertilized eggs from a mating that took place within a fig, then find another fig of the same species, enter it, lay eggs, and die.

Her offspring will then hatch, develop, and mate within the fig. Because each species of fig wasp reproduces only in its particular species fig, each wasp will have the opportunity to mate only with other wasps of its own species. This would be considered _____.
   a. ecological isolation
   b. geographical isolation

7. A likely group of animals in which to look for cases of sympatric speciation would be
   a. birds.
   b. parasitic intestinal worms.
   c. bats.

8. Which of the following ecological barriers was a major factor in the evolution of *Rhagoletis pomonella?*
   a. a river separating the apple orchard and the forest with hawthorn trees
   b. a difference in canopy height between the two trees
   c. the greater appeal to females of males that smell like apples
   d. the preference of females for ovipositing (laying eggs) on different fruits

9. A new variety of *Rhagoletis pomonella* has been discovered infesting cherry trees. What should cherry-producing farmers be told?
   a. The new flies arose on an island having only cherry trees, separated from other sources of fruit.
   b. The infestation is only temporary, and the flies will eventually switch back to hawthorn.
   c. The flies have switched from apple trees and may be evolving into a new "cherry-only" species by sympatric speciation.

10. A mechanism by which many plants but few animal species have evolved sympatrically is through
    a. polyploidy.
    b. genetic drift.
    c. gene flow.
    d. aneuploidy.

11. Genetic divergence is required for speciation to occur, but how can speciation be guaranteed?
    a. There are mechanisms preventing interbreeding between developing species.
    b. Individuals from developing species are kept from being able to mate with each other.

c. Offspring produced from the mating of individuals from developing species are unable to pass along their genes to a subsequent generation.

d. all of the above

12. Which of the following is considered a requirement for speciation to occur?

 a. Populations must be isolated (geographically or in some other manner) from one another.

 b. Isolated populations must become genetically distinct from one another.

 c. Exchanges of genetic information must be restricted between populations.

 d. All of the above are correct.

13. Genetic drift can contribute to allopatric speciation if

 a. the original population is small enough.

 b. one of the separated populations is small enough.

 c. one population undergoes a bottleneck just before being reunited with the other.

14. When a species has no surviving members, it is said to be _____.

 a. temporally isolated

 b. extinct

 c. ecologically isolated

 d. adaptively radiating

15. Which of the following events is the leading cause of extinction?

 a. localized species distribution

 b. species overspecialization

 c. competition among species

 d. habitat change

QUIZ 2

1. Which of the following is most characteristic of populations of *different* species?

 a. Members of the two populations resemble each other.

 b. Members of the two populations can be distinguished by their appearance.

 c. The two populations are geographically separated from each other.

 d. The two populations are adapted to different habitats.

 e. A fertile female from one population mated with a fertile male from the other population produces no offspring.

2. Under the biological-species concept, the main criterion for identifying a species is

 a. anatomical distinctiveness.

 b. behavioral distinctiveness.

 c. geographic isolation.

 d. reproductive isolation.

3. An ecological barrier prevents populations from interacting during critical points in their life cycle, such as

 a. feeding.

 b. mating.

 c. overwintering.

4. Incompatibilities that prevent the formation of vigorous, fertile hybrids between species are called

 a. premating isolation mechanisms.

 b. postmating isolation mechanisms.

 c. geographical isolation.

 d. genetic divergence.

5. Which of the following is a premating reproductive-isolating mechanism?

 a. ecological isolation

 b. temporal isolation

 c. behavioral isolation

 d. mechanical incompatibility

 e. all of the above

6. Bishop pines and Monterey pines coexist in nature. In the laboratory, they produce fertile hybrids; but in the wild, they do not interbreed. This is because each species releases pollen at different times of the year. This is an example of _____.

 a. temporal isolation

 b. behavioral isolation

 c. mechanical isolation

 d. geographical isolation

7. Having multiple sets of chromosomes beyond the diploid number is known as

 a. polyploidy.

 b. genetic drift.

 c. gene flow.

 d. aneuploidy.

8. Which of the following is NOT an example of speciation?

 a. A small group from a large mainland population colonizes a remote island.

 b. A river that has long divided two populations of mice is diverted by an earthquake, and the two mouse populations come into contact and breed. The hybrid offspring, however, are sterile.

 c. In a bird population, there is disruptive selection for habitat: one group adapts to the treetops, while another adapts to the lower branches and ground. The two groups rarely interbreed; but when they do, the hybrid offspring do not live long because they have a mixture of both kinds of adaptations and are not adapted to either habitat.

d. Over a period of several million years, a deerlike species evolves from being very small and feeding on grasses and small shrubs to being much larger and feeding on the lower branches of trees.

e. Due to meiotic error, a diploid plant capable of self-fertilization produces a self-fertilizing tetraploid offspring.

9. Populations of two species living in the same areas (e.g., chorus frogs and wood frogs living in the same ponds of Ohio woodlots) are said to be _____.
   a. allopatric
   b. sympatric
   c. convergent
   d. divergent

10. The rapid speciation of Darwin's finches on the Galapagos Islands is an example of _____.
    a. coevolution
    b. adaptive radiation
    c. convergent evolution

11. Which of the following is the first step in the process of allopatric speciation?
    a. interspecies contact
    b. geographic isolation
    c. independent evolution of two species
    d. reproductive isolation

12. Which of the following is (are) likely to promote sympatric speciation?
    a. gene flow
    b. geographic isolation
    c. ecological isolation
    d. chromosomal aberrations
    e. (c) and (d)

13. If a haploid egg from a diploid plant is fertilized by a haploid sperm from a diploid plant, and the fertilized egg duplicates its chromosomes but does not divide into two cells, the resulting cell may then divide normally, producing an individual plant that will be _____.
    a. haploid
    b. diploid
    c. triploid
    d. tetraploid
    e. none of the above

14. Which of the following may cause a species to become extinct?
    a. habitat encroachment (e.g., urbanization)
    b. seasonal changes in the weather
    c. an inability to successfully compete for limited resources
    d. (a) and (c)

15. Which species is at LEAST risk of extinction?
    a. The species that has many geographically isolated populations, all of them small.
    b. The species that is composed of one large, continuous, genetically variable population.
    c. The species that lives only in a tree that is itself endangered.
    d. The species' major food source is an insect population that is declining because of pesticide use.
    e. A native plant species that lives where a newly introduced nonnative plant has adapted and grows more quickly.

ANSWER KEY

Quiz 1

1. d
2. a
3. d
4. a
5. a
6. a
7. b
8. d

9. c
10. a
11. d
12. d
13. b
14. b
15. d

Quiz 2

1. e
2. d
3. b
4. b
5. e
6. a
7. a
8. a

9. b
10. b
11. b
12. e
13. d
14. d
15. b

CHAPTER 17: THE HISTORY OF LIFE

OUTLINE

Section 17.1 How Did Life Begin?

- Prior to life on Earth, organic molecules (e.g., nucleic acids, amino acids, and lipids) may have formed when lightning, ultraviolet light, and heat reacted with components of the prebiotic atmosphere (**Figure 17-2**).
- RNA many have been the earliest form of nucleic acid and may have functioned as an enzyme (**ribozyme**), replicating itself from nucleotides available in the waters of early Earth.
- Aggregations of proteins and lipids may have formed vesicles that surrounded ribozymes, forming a semblance of a living cell (a **protocell**).

Section 17.2 What Were the Earliest Organisms Like?

- Early Precambrian life consisted of prokaryotic bacteria that were likely anaerobic (because of the lack of available atmospheric oxygen) and absorbed required organic molecules from the surrounding environment. Over time, depletion of available organics likely explained the development of photosynthesis, which allowed some bacteria to synthesize their own food using available inorganic molecules.
- Photosynthetic activity resulted in significant accumulations of atmospheric oxygen 2.2 billion years ago. The availability of oxygen likely resulted in the development of aerobic metabolism, which produces more cellular energy than anaerobic mechanisms.
- Eukaryotic cells first evolved 1.7 billion years ago. Membranous organelles may have been derived from infolding of the plasma membrane. The **endosymbiont hypothesis** suggests that mitochondria and chloroplasts evolved from bacteria engulfed by predatory cells (**Figures 17-4** and **17-5**).

Section 17.3 What Were the Earliest Multicellular Organisms Like?

- Multicellular life evolved in the form of eukaryotic algae 1.2 billion years ago. The algae could anchor themselves along the shore and take advantage of the abundant light and nutrients. Larger size also protected algae from predators.
- Multicellular invertebrate animals first appeared about 610 million years ago. For these animals, multicellularity allowed faster locomotion, which imparted more effective prey capture and escape from predation. These behaviors established environmental pressures for faster locomotion, as well as sensory and intellectual development (**Figure 17-6**).

Section 17.4 How Did Life Invade the Land?

- Land plants first evolved 400 million years ago. Terrestrial existence required the development of body support, as well as adaptations to prevent desiccation and allow reproduction out of water. However, terrestrial existence allowed access to abundant light in the absence of predators.
- **Arthropods** were the first animals to move onto land (occurring soon after the evolution of land plants), primarily because of the suitability of their **exoskeleton** for a terrestrial existence.
- The first terrestrial vertebrates evolved from **lobefin fishes**, which possessed fleshy fins and a primitive lung that were used to facilitate life on land (**Figure 17-8**).
- The development of improved legs and lungs allowed the evolution of **amphibians** from lobefin fishes 350 million years ago. Although better suited for terrestrial life, amphibians needed to remain close to water to reproduce and sustain cutaneous respiration.
- **Reptiles** evolved from amphibians and were less dependent on the proximity of water, as a result of a variety of adaptations, including waterproof scales and eggs and better developed lungs (**Figure 17-9**).
- **Mammals** and **birds** evolved from separate reptile ancestors, developing body insulation in the form of hair and feathers respectively.

Section 17.5 What Role Has Extinction Played in the History of Life?

- New species evolved and existing species became extinct throughout the history of life.
- **Mass extinctions** periodically occur, resulting in the extinction of many species in a short period of time (**Figure 17-10**). Mass extinctions can occur from climate change (i.e., continental drift from **plate tectonics**, **Figure 17-11**) and catastrophic events.

Section 17.6 How Did Humans Evolve?

- Apes and humans evolved from tree-dwelling **primate** mammals (**Figure 17-12**) with binocular vision, grasping hands, and a relatively large brain.
- The **hominid** lineage diverged from that of the apes between 5 and 8 million years ago; the oldest hominid fossils were found in Africa (**Figure 17-13**).
- The first well-known hominids, the australopithecines, lived 4 million years ago. They could stand and walk upright and had larger brains than their ancestors.
- The genus *Homo* diverged from the australopithecines 2.5 million years ago and was accompanied by advances in tool technology (**Figures 17-14** and **17-15**). There are several lineages of the genus *Homo* (**Figure 17-18**), which include the Neanderthals (*H. neanderthalensis*) and modern humans (*H. sapiens*).

LEARNING OBJECTIVES

Section 17.1 How Did Life Begin?

- Discuss the importance of the early atmosphere's lack of oxygen on organic development.
- Discuss the components necessary for the spontaneous formation of complex molecules.
- Explain how clay aids in the production of complex organic molecules.
- Discuss the various functions of ribozymes.

Section 17.2 What Were the Earliest Organisms Like?

- Discuss the production and accumulation of oxygen in the atmosphere and its impact on early organisms.
- Explain the endosymbiotic theory in terms of mitochondria and chloroplasts.
- Discuss the scientific evidence supporting the endosymbiotic theory.

Section 17.3 What Were the Earliest Multicellular Organisms Like?

- Explain the role diffusion plays in determining the size of cells.
- Discuss the advantages to an organism to develop multicellularity.
- Discuss the succession of evolving traits in animal evolution.

Section 17.4 How Did Life Invade the Land?

- Explain the obstacles faced by organisms as they moved from water to a land habitat.
- Discuss the benefits provided by colonizing land.
- Discuss the adaptations necessary for plants to reproduce on dry land.
- Explain the advantages of insect/animal pollination over wind pollination.
- Explain why arthropods were the first to colonize land.
- Discuss the attributes of lobefishes that made them capable of living on land.
- Discuss the succession of animals from lobefishes to mammals.

Section 17.5 What Role Has Extinction Played in the History of Life?

- Discuss the role of climate in mass extinctions.
- Discuss the role of asteroids in mass extinctions.

Section 17.6 How Did Humans Evolve?

- Discuss the key features of primates.
- Discuss the advantages of walking upright.
- Discuss the differences in brain size, tool usage, and culture among early hominids and modern man.

QUIZ 1

1. Which of the following was NOT a common component of Earth's early atmosphere?
   a. hydrogen
   b. oxygen
   c. methane
   d. ammonia
   e. carbon dioxide

2. Which of the following included the "age of dinosaurs"?
   a. Precambrian
   b. Mesozoic
   c. Paleozoic
   d. Cenozoic

3. The earliest living organisms were
   a. multicellular.
   b. eukaryotes.
   c. prokaryotes.
   d. photosynthesizers.

4. The MOST fundamental difference between prokaryotes and eukaryotes is that eukaryotes have

   _____.
   a. a cell wall
   b. a plasma membrane
   c. a membrane-bound nucleus
   d. genetic material
   e. multiple cells

5. The first multicellular organisms were thought to be
   a. arthropods.
   b. conifers.
   c. cnidarians.
   d. algae.

6. The exoskeleton of early, marine-dwelling arthropods can be considered a preadaptation for life on land because the shell
   a. can support an animal's weight against gravity
   b. allows a wide diversity of body types
   c. resists drying
   d. absorbs light
   e. (a) and (c)

7. The earliest animals to have an amniotic egg, allowing eggs to develop out of water, were
   a. lobefinned fish.
   b. amphibians.
   c. reptiles.
   d. birds.

8. An advantage that land plants have over aquatic plants is that land plants
   a. do not need elaborate systems of support.
   b. are capable of more rapid photosynthesis.
   c. have better access to water.
   d. can reproduce more easily.

9. The earliest land plants were restricted to wetlands habitants because they
   a. were single-celled.
   b. did not have a root system.
   c. could not stay upright.
   d. had swimming sperm rather than sperm encased in pollen grains.

10. Present day coal deposits are actually remains of _____ that lived during the _____.
    a. marine invertebrates, Paleozoic
    b. plants, Carboniferous
    c. reptiles, Jurassic
    d. mammoths, Cenozoic

11. The supercontinent of Laurasia gave rise to the current continents of
    a. North America, South America, and Antarctica.
    b. Africa, Australia, Europe, and Asia.
    c. North America, Europe, and Asia.
    d. South America, Africa, Antarctica, and Australia.

12. Each continent began to develop its own unique flora and fauna about 135 million years ago, when
    a. Pangaea had not yet formed.
    b. Pangaea had unified as one.
    c. Laurasia and Gondwanaland were formed.
    d. Laurasia and Gondwanaland broke up.
    e. the continents reached their current configuration.

13. Which statement about primates is true?
    a. At least two groups of primates have evolved to lose their tree-dwelling existence.
    b. At least one group of primates has evolved to lose opposable digits.
    c. At least one group of primates has evolved to lose binocular vision.
    d. All modern primates are strictly herbivorous.

14. Given the following list, which is the correct numbered sequence from earliest to most recent hominid development?
    1. bipedalism
    2. cave paintings

3. first stone tools made
4. first migration out of Africa
5. buried dead and "religious" rites
   a. 1, 2, 3, 4, 5
   b. 3, 1, 2, 4, 5
   c. 1, 3, 4, 5, 2
   d. 3, 2, 1, 5, 4
   e. 1, 4, 3, 5, 2

15. Which of the following features distinguishes the hominid line from the lineage of apes?
    a. binocular vision
    b. large brain
    c. hands with opposable thumbs
    d. dependence upon an upright posture for locomotion

QUIZ 2

1. Which of the following would have prevented the development of conditions favorable for the evolution of life after Earth was formed?
   a. complex organic molecules
   b. energy in the form of lightning and energy from the sun
   c. unbound, or free, oxygen molecules
   d. carbon dioxide and sulfur dioxide

2. What organic macromolecule do scientists believe was the first one able to generate more copies of itself during prebiotic conditions?
   a. DNA
   b. RNA
   c. protein
   d. none of the above

3. The reason that both mitochondria and chloroplasts have a double membrane is that the inner membrane belonged to the original bacteria and the outer membrane
   a. was formed from a predatory cell membrane during phagocytosis.
   b. arose from the bacteria being surrounded by endoplasmic reticulum.
   c. was created by a mutation that later duplicated the bacterial cell membrane.
   d. was synthesized by the Golgi apparatus.

4. What experimental evidence suggests that organelles of eukaryotic cells formed from endosymbiotic relationships between bacterial cells?
   a. Both chloroplasts and mitochondria contain DNA that is distinct from that found in the nucleus of the eukaryotic cell.
   b. There are similarities in both structure and function between the cilia and flagella of eukaryotes and the coiled/spiral form of bacteria.
   c. There are present-day examples in which a eukaryotic organism contains either a population of bacteria or a different form of algae.
   d. All of the above are correct.
   e. None of the above is correct.

5. Life may have been present on Earth as long ago as nearly
   a. 4 billion years.
   b. 4 million years.

   c. 400,000 years.
   d. 40,000 years.
   e. 4000 years.

6. Which of the following characteristics is believed to be an evolutionary advantage of multicellular eukaryotic organisms?
   a. Multicellular organisms are easily ingested by single-celled predators.
   b. Multicellular organisms were able to let certain cells specialize to carry out functions that single-celled organisms could not.
   c. Multicellularity allowed cells to remain small and able to exchange carbon dioxide and oxygen efficiently, despite increasing the size of the organism.
   d. (b) and (c) are correct.
   e. None of the above is correct.

7. If a species first evolved in the seas, which of the following would NOT be an obstacle to its effort to colonize land?
   a. dehydration of cells and tissues
   b. the external skeleton of arthropods
   c. exchange of gametes between individuals
   d. locomotion

8. The first animals to "invade" the land were probably _____.
   a. arthropods
   b. fish
   c. amphibians
   d. reptiles

9. Which of the following was an evolutionary advancement that allowed reptiles to exploit life on land fully?
   a. improved lungs
   b. internal fertilization
   c. shelled, waterproof eggs
   d. scaly, waterproof skin
   e. all of the above

10. The reptiles gave rise to _____.
    a. fish
    b. amphibians
    c. birds
    d. mammals
    e. (c) and (d)

11. Earth is composed of a
    a. solid surface crust and a hollow, gaseous interior.
    b. solid surface crust and a liquid metal interior.
    c. liquid metal surface crust and a solid interior.

12. Earth's continents were unified into a single super-continent called
    a. Pangaea.
    b. Gondwanaland.
    c. Laurasia.
    d. Gaia.
    e. Atlantis.

13. Which statement BEST describes the pattern of extinction throughout biological history?
    a. The rate of extinction has steadily increased over the entire history of life.
    b. The rate of extinction has steadily decreased over the entire history of life.
    c. There has been very little extinction except during mass extinction events.
    d. There has been a relatively constant turnover of species with occasional mass extinction events.

14. Which of the following was a likely characteristic of the early primates?
    a. color vision
    b. long, grasping fingers
    c. a proportionally larger brain
    d. forward-facing eyes with binocular vision
    e. all of the above

15. The hominids belonging to the genus *Australopithecus* first evolved in _____.
    a. Asia
    b. Africa
    c. Australia
    d. Europe

ANSWER KEY

Quiz 1

1. b
2. b
3. c
4. c
5. d
6. e
7. c
8. b
9. d
10. b
11. c
12. d
13. a
14. c
15. d

Quiz 2

1. c
2. b
3. a
4. d
5. a
6. d
7. b
8. a
9. e
10. e
11. b
12. a
13. d
14. e
15. b

CHAPTER 18: SYSTEMATICS: SEEKING ORDER AMIDST DIVERSITY

OUTLINE

Section 18.1 How Are Organisms Named and Classified?

- Systematists places organisms into hierarchical categories based on features that reveal their evolutionary relationships. In order of increasing inclusiveness, these categories are **domain**, **kingdom**, **phylum**, **class**, **order**, **family**, **genus**, and **species** (**Table 18-1**).
- Each species is identified by a **scientific name**, composed of its genus name and species name. The first letter of the genus name is always capitalized, and the entire scientific name is either italicized or underlined (e.g., *Sialia sialis*).

Section 18.2 What Are the Domains of Life?

- All life is classified into three domains that represent the main branches of the tree of life (**Figure 18-5**). These domains are **Bacteria**, **Archaea**, and **Eukarya** (**Figure 18-4**).
- Domains are composed of **kingdoms** of organisms, but kingdom-level classification is currently in transition. Among the Eukarya, commonly accepted kingdoms include the Fungi, Plantae, and Animalia (**Figure 18-6**).

Section 18.3 Why Do Classifications Change?

- Classification schemes are revised as new information is discovered.
- The biological species concept, although useful, is not applicable to some organisms (primarily asexually reproducing organisms), which makes them difficult to classify. In these cases, alternative definitions of species can be useful.

Section 18.4 How Many Species Exist?

- **Biodiversity** refers to the total range of species diversity on Earth.
- The total number of named species is currently 1.5 million, with 700 to 10,000 new species named annually. Scientists believe there may be as many as 100 million species on Earth.

LEARNING OBJECTIVES

Section 18.1 How Are Organisms Named and Classified?

- Explain the benefits of scientific nomenclature.
- Discuss the roles of anatomy and DNA in phylogenetic hierarchies.

Section 18.2 What Are the Domains of Life?

- Discuss the basis of the three-domain system.

Section 18.3 Why Do Classifications Change?

- Discuss the different ways to determine species identity.

Section 18.4 How Many Species Exist?

● Discuss the discrepancy between named organisms and those yet to be described.

QUIZ 1

1. Anatomical features of species may not always be useful for determining species relationships because of _____.
   a. convergent evolution
   b. homologous structures
   c. adaptation
   d. common ancestry

2. What is the most inclusive of the major taxonomic categories?
   a. genus
   b. order
   c. phylum
   d. kingdom
   e. domain

3. What is the least inclusive of the major taxonomic categories?
   a. species
   b. order
   c. phylum
   d. division
   e. class

4. In the following sequence, what is missing from the hierarchy of major taxonomic categories (list in order): species—_____—family—_____—class.
   a. genus, domain
   b. genus, order
   c. kingdom, order
   d. order, genus
   e. order, division
   f. phylum, genus

5. The more taxonomic categories two organisms share, the more closely related those organisms are in an evolutionary sense. Which scientist's work led to this insight?
   a. Aristotle
   b. Darwin
   c. Linnaeus
   d. Whittaker
   e. Woese

6. Pretend that you could divide a funnel into seven horizontal layers and then you put all of the taxonomic categories into the funnel. Which taxonomic categories would trickle out of the bottom of the funnel, one at a time?
   a. kingdoms and phyla
   b. phyla and classes
   c. classes and families
   d. families and genera
   e. genus and species

7. It is possible to imagine the various levels of taxonomic classification as a "family tree" for an organism. If the kingdom is analogous to the trunk of the tree, which taxonomic category would be analogous to the large limbs coming off that trunk?
   a. class
   b. family
   c. order
   d. phylum
   e. subfamily

8. Molecular genetics helps us distinguish and evaluate the degree of evolutionary relatedness among species by allowing us to
   a. compare nucleotide sequences of DNA from the same chromosomal regions from different species.
   b. compare DNA sequences from different chromosomal regions between two closely related species.
   c. (a) and (b) are both correct.
   d. Neither (a) nor (b) is correct.

9. Which kingdom comprises eukaryotic organisms that are multicellular and heterotrophic and do not possess cell walls?
   a. Animalia
   b. Fungi
   c. Archaea
   d. Plantae
   e. Protista

10. What is a domain?
   a. the broadest of the three taxonomic categories
   b. the taxonomic category that contains insects, birds, and mammals
   c. the taxonomic category that contains all living species of humans
   d. a synonym for the taxonomic category, kingdom

11. Humans belong to the domain
   a. Primates.
   b. Hominidae.
   c. *Homo*.
   d. Eukarya.
   e. Mammalia.

12. Changes at the domain or kingdom level of classification _____.
   a. never occur
   b. rarely occur
   c. occur about every five years
   d. occur very frequently

13. The phylogenetic species concept stresses _____ as the criterion for assigning individuals to the same species.
 a. anatomical similarity
 b. recent common ancestry
 c. potential for interbreeding
 d. similarity in behavior

14. There are approximately 1.5 million named species today. Of these, approximately 5% are prokaryotes and protists. Scientists believe that
 a. this is a good representation of the true percentage of all species that are either prokaryotes or protists.
 b. this is an overestimation of the true percentage of all species that are either prokaryotes or protists.
 c. this is an underestimation of the true percentage of all species that are either prokaryotes or protists.
 d. all possible protist and prokaryote species have been discovered already.

15. Which one of the following habitats appears to have the greatest number of species?
 a. seafloor
 b. deserts
 c. tropical rain forests
 d. grasslands

QUIZ 2

1. The scientific name is composed of two classification categories, the
 a. Domain and Kingdom.
 b. antonym and homonym.
 c. genus and species.

2. Which of these scientific names of species is correctly written?
 a. *Aneides Aeneus* (the green salamander)
 b. *Crotalus horridus* (the timber rattlesnake)
 c. *falco peregrinus* (the peregrine falcon)
 d. Marmota monax (the woodchuck)
 e. *Salmo* (the rainbow trout)

3. The science of reconstructing evolutionary history is
 a. ethology.
 b. archaeology.
 c. anthropology.
 d. systematics.
 e. systems analysis.

4. Classification categories are arranged into a _____, in which each successive category is increasingly narrow and specifies a group whose common ancestor is increasingly recent.
 a. ladder
 b. hierarchy
 c. stairway
 d. checkerboard

5. The correct sequence of the taxonomic hierarchy is
 a. Domain, Kingdom, Class, Phylum, Order, Family, Genus, Species.
 b. Domain, Kingdom, Phylum, Order, Class, Family, Genus, Species.
 c. Domain, Kingdom, Phylum, Class, Family, Order, Species, Genus.
 d. Kingdom, Domain, Order, Phylum, Class, Family, Genus, Species.
 e. Domain, Kingdom, Phylum, Class, Order, Family, Genus, Species.

6. Humans belong to the Order
 a. Primates.
 b. Hominidae.
 c. *Homo*.
 d. Mammalia.

7. Which of the following criteria could NOT be used to determine how closely related two types of organisms are?
 a. similarities in the presence and relative abundance of specific molecules
 b. DNA sequence
 c. the presence of homologous structures
 d. developmental stages
 e. occurrence of both organisms in the same habitat

8. Which of the following pairs is the most distantly related?
 a. archaea and bacteria
 b. protists and fungi
 c. plants and animals
 d. fish and starfish
 e. fungi and plants

9. An organism is described to you as having many nuclei-containing cells, each surrounded by a cell wall of chitin, and as absorbing its food. In which of the following taxonomic groups would you place it?
 a. Plantae
 b. Protista
 c. Animalia
 d. Fungi
 e. Archaea

10. Which taxonomic group contains eukaryotic organisms that are typically unicellular?
    a. Animalia
    b. Fungi
    c. Archaea
    d. Plantae
    e. Protista

11. Which kingdom contains eukaryotic organisms that are multicellular and autotrophic and possess cell walls?
    a. Animalia
    b. Fungi
    c. Bacteria
    d. Plantae
    e. Protista

12. A major difficulty in using the biological species concept is that
    a. related organisms resemble each other too much to distinguish.
    b. asexual organisms do not fulfill the criterion of interbreeding.
    c. most organisms have the capacity to interbreed with members of other species.
    d. None of the above is correct.

13. Which of the following is NOT used by taxonomists to distinguish one species from another?
    a. genetic characteristics
    b. similar features resulting from convergent evolution acting on distantly related species

    c. distinguishing features seen in organisms during its development
    d. external anatomical features

14. Why can classification schemes (taxonomies) change from year to year?
    a. A population of individuals of one species evolves into a new species.
    b. New species are discovered.
    c. New data about previously described species are discovered.
    d. All of the above are correct.
    e. (b) and (c) are correct.

15. Which of the following statements about biodiversity is (are) TRUE?
    a. Bottled water is more biodiverse than pond water.
    b. Tropical forests are more biodiverse than temperate forests.
    c. Arthropods exhibit more biodiversity than mammals.
    d. (b) and (c) are correct.

ANSWER KEY

Quiz 1

1. a		9. a	
2. e		10. a	
3. a		11. d	
4. b		12. b	
5. b		13. b	
6. e		14. c	
7. d		15. c	
8. a			

Quiz 2

1. c		9. d	
2. b		10. e	
3. d		11. d	
4. b		12. b	
5. e		13. b	
6. a		14. e	
7. e		15. d	
8. a			

CHAPTER 19: THE DIVERSITY
OF PROKARYOTES AND VIRUSES

OUTLINE

Section 19.1 Which Organisms Make Up the Prokaryotic Domains—Bacteria and Archaea?

- Bacteria and Archaea are distinguished by fundamental differences in their structural and biochemical features. For example, Bacteria have peptidoglycan in their cell walls (Archaeans do not), and both groups differ in the composition of their plasma membranes, ribosomes, and RNA polymerases.
- Prokaryotic classification within domains is difficult and occurs on the basis of shape, means of locomotion, pigmentation, nutrient requirements, appearance of colonies, staining properties, and genetic similarity.
- Prokaryotes can be found in a wide range of sizes and have common characteristic shapes, including spherical, rodlike, and corkscrew (**Figure 19-1**).

Section 19.2 How Do Prokaryotes Survive and Reproduce?

- Prokaryotes have different ranges of mobility. Some adhere to surfaces or drift in their liquid surroundings, but others actively move by the use of flagella (**Figure 19-2**).
- Some bacterial cell walls are surrounded by a protective slime (**biofilm**) that helps them adhere to surfaces (**Figure 19-3**).
- Under harsh environmental conditions, many rod-shaped bacteria can form **endospores** that allow genetic material (and some enzymes) to survive for long periods of time in metabolic dormancy (**Figure 19-4**).
- Prokaryotes are specialized for survival in specific habitats (**Figure 19-5**) and exhibit many different methods of energy acquisition, including anaerobic, aerobic, chemosynthetic, and photosynthetic metabolisms (**Figure 19-6**).
- Prokaryotes typically reproduce by asexual binary fission (**Figure 19-7**) but may exchange genetic material by a process called **conjugation**. During this process, circular DNA **plasmids** are transferred from donor to recipient prokaryotes (**Figure 19-8**).

Section 19.3 How Do Prokaryotes Affect Humans and Other Eukaryotes?

- Prokaryotes play a number of important roles in the lives of eukaryotes. Many herbivorous animal digestive systems could not extract nutrients from plant matter without the help of bacteria to break down plant cellulose. **Nitrogen-fixing bacteria** provide plants with a source of soil nitrogen by converting atmospheric nitrogen into ammonia (**Figure 19-19**). Other prokaryotes decompose dead organic matter, acting as nature's recyclers, while others can break down pollution resulting from human activity.
- Soma bacteria are **pathogenic** and can cause a number of human diseases, including tetanus, the plague, strep throat, and pneumonia. The majority of bacteria, however, are harmless.

Section 19.4 What Are Viruses, Viroids, and Prions?

- **Viruses** are small, nonliving, host-specific parasites consisting of a protein coat that surrounds genetic material (**Figure 19-11**). They reproduce only by invading a host cell and, by using the cell's own enzymes, directing it to make more viruses, which exit the cell when it ruptures (**Figure 19-12**). Many viruses are pathogenic.
- **Viroids** are circular strands of RNA that enter the cell nucleus and direct new viroid production. Viroids are known to affect only plant cells.
- **Prions** are mutated proteins that may act enzymatically to form more prions, thus disrupting cell function (**Figure 19-14**). Prions have been implicated in a number of neurological diseases.

LEARNING OBJECTIVES

Section 19.1 Which Organisms Make Up the Prokaryotic Domains—Bacteria and Archaea?

- Describe the differences between bacteria and archaea.

Section 19.2 How Do Prokaryotes Survive and Reproduce?

- Describe the various bacterial attributes that allow them to thrive in virtually all environments.
- Describe the different prokaryotic metabolisms.

Section 19.3 How Do Prokaryotes Affect Humans and Other Eukaryotes?

- Discuss the positive roles prokaryotes play in animal and plant metabolism.
- Describe the environmental benefits provided by prokaryotes.
- Describe the different diseases caused by bacteria.

Section 19.4 What Are Viruses, Viroids, and Prions?

- Discuss the differences between viruses and prokaryotes.
- Describe the basic viral life cycle.
- Discuss the difficulties in developing antiviral medications.
- Discuss the difference between prions and other infectious agents.

QUIZ 1

1. Prokaryotes have
   a. no true nucleus.
   b. no membranous organelles.
   c. a large, circular chromosome.
   d. all of the above.

2. Peptidoglycan is a component of the cell wall of
   a. bacteria
   b. archaea
   c. both of the above
   d. neither of the above

3. Classification of prokaryotes may use many kinds of traits, including which of the following?
   a. cell shape
   b. means of locomotion
   c. nutrient sources
   d. staining properties
   e. all of the above

4. The small circles of DNA found only in prokaryotes are known as
   a. genes.
   b. chromosomes.
   c. plasmids.

5. During bacterial conjugation, the transferred item is a
   a. gene.
   b. chromosome.
   c. plasmid.

   d. lipid.
   e. protein.

6. Which of the following traits allows some bacteria to survive extreme conditions for millions of years?
   a. aerobic respiration
   b. endospore formation
   c. conjugation
   d. binary fission

7. The simple form of cell division by which prokaryotic cells reproduce is called _____.
   a. binary fission
   b. conjugation
   c. endospore formation
   d. mitosis

8. Which of the following pairs of organism and disease is INCORRECT?
   a. virus: AIDS
   b. prion: kuru
   c. bacterium: syphilis
   d. archaean: gonorrhea

9. Which of the following enables plants to obtain a usable form of nitrogen?
   a. bacteria
   b. viruses
   c. prions
   d. zooflagellates

10. The majority of bacteria are
    a. pathogenic.
    b. photosynthetic.
    c. harmless.
    d. eukaryotic.
    e. harmful.

11. What are viruses, viroids, and prions?
    a. nonliving, infectious agents or pathogens that can cause disease in organisms
    b. small sequences of nucleotides or amino acids that are devoid of their own cell membrane
    c. substances that can cause disease only after entering a host cell and taking over the nuclear machinery
    d. all of the above

12. Which of the following is TRUE?
    a. Viruses cannot reproduce outside a host cell.
    b. Prions are infectious proteins.
    c. Viroids lack a protein coat.
    d. Some viruses cause cancer.
    e. All of the above are correct.

13. Viruses that attack bacteria are called
    a. prions.
    b. bacteriophages.
    c. herpes viruses.
    d. mosaic viruses.
    e. viroids.

14. Which components make up a virus?
    a. protein particles only
    b. RNA only
    c. DNA or RNA and a protein coat
    d. DNA and RNA and a protein coat
    e. DNA in a nucleus, RNA, ribosomes, plasma membrane, and cell wall

15. Viruses make new viral copies (reproduce) in
    a. water.
    b. the soil.
    c. a host cell.
    d. the air.

QUIZ 2

1. Bacteria and archaea differ in the structure and composition of their
   a. plasma membrane.
   b. ribosomes.
   c. RNA polymerases.
   d. all of the above.

2. Following classification into unique domains, bacteria and archaea have proven
   a. easy to classify into kingdoms because of their size
   b. easy to classify into kingdoms because of the biochemical differences
   c. difficult to classify into kingdoms because of their size
   d. difficult to classify into kingdoms because of the biochemical differences

3. Gram stain
   a. can be used to help differentiate some classes of bacteria.
   b. turns gram–positive bacteria green.
   c. gives the researcher information about the construction of the cell wall.
   d. all of the above
   e. some but not all of the above

4. At the end of conjugation
   a. one cell has lost a plasmid, while the other has gained one.
   b. one cell has lost a chromosome, while the other has gained one.
   c. the two cells have had an equal exchange of plasmids.
   d. one cell has given a copy of a plasmid to another, while keeping one copy for itself.

5. After the plasmid is transferred, it must be
   a. excised of its introns.
   b. excised of its exons.
   c. transcribed into a loop of complementary RNA.
   d. replicated to form double-stranded DNA.

6. Sexual reproduction performed by eukaryotes is similar to bacterial conjugation because
   a. two haploid cells fuse.
   b. genetic variability increases.
   c. crossing over occurs.
   d. genetically different diploid offspring form.

7. Hospitals must sterilize surgical instruments at very high temperature and pressure because some bacteria can survive harsh conditions by making _____.
   a. zygotes
   b. endospores
   c. seeds
   d. fruiting bodies

8. Which of these is a bacterial disease?
   a. influenza
   b. common cold
   c. AIDS
   d. tetanus

9. Which of the following is (are) a producer in its ecosystem because of the photosynthesis it performs?
   a. prions
   b. viruses
   c. cyanobacteria
   d. all of the above

10. Which of the following role(s) do bacteria play in the environment?
    a. producers
    b. decomposers
    c. partners in mutualistic symbiosis
    d. pathogens
    e. all of the above

11. HIV encounters its host cells
    a. via direct contact with another person.
    b. by traveling through the blood after entering the body via contact with blood or body fluids.
    c. via inhalation from the air.

12. The host cell in which the HIV virus reproduces is a
    a. nerve cell.
    b. red blood cell.
    c. helper T cell.
    d. skin cell.

13. After a retrovirus enters its host cell
    a. reverse transcription creates a strand of DNA complementary to the viral RNA.
    b. transcription creates mRNA from viral RNA.

c. viral DNA is translated into new viral proteins, such as reverse transcriptase.
d. viral DNA and new viral proteins are packaged into pieces of the cell membrane when the new viruses bud off the cell.

14. After reverse transcription creates viral DNA from the viral RNA, the viral DNA
    a. is translated into new viral proteins.
    b. is packaged together with new viral proteins to form new viruses.
    c. enters the host cell's nucleus and is integrated into the host's chromosome.
    d. is used as a template to make new host cell DNA.
    e. remains in the cytoplasm where new viruses are assembled.

15. After viral DNA and viral envelope proteins are assembled into new viruses, they exit the host cell via
    a. transcription.
    b. exocytosis.
    c. mitosis.
    d. endocytosis.

ANSWER KEY

Quiz 1

1. d	9. a
2. a	10. c
3. e	11. d
4. c	12. e
5. c	13. b
6. b	14. c
7. a	15. c
8. d	

Quiz 2

1. d	9. c
2. c	10. c
3. e	11. b
4. d	12. c
5. d	13. a
6. b	14. c
7. b	15. b
8. d	

CHAPTER 20: THE DIVERSITY OF PROTISTS

OUTLINE

Section 20.1 What Are Protists?

- Protists are a generalized group of eukaryotes that are not plants, animals, or fungi. Although most protists are single celled, some form colonies and some are multicellular.
- Protists exhibit several modes of nutrition, including heterotrophic use of **pseudopods** (**Figure 20-1**), absorptive feeders, and photosynthesis.
- Protists reproduce through both asexual and sexual methods, suggesting that sexual reproduction first arose in eukaryotes (**Figure 20-2**).
- Although protists make up an important component of marine ecosystems, there are those that parasitize humans, cause plant diseases, and release toxins into the environment.

Section 20.2 What Are the Major Groups of Protists?

- **Excavates** are protists that lack nuclei. They include the **diplomonads**, which have two nuclei (**Figure 20-3**), and **parabasalids**, which all live inside animals (**Figure 20-4**).
- **Eulenozoans** have distinctively shaped mitochondria. They include the **euglenids**, which lack a rigid covering and swim by flagella (**Figure 20-5**), and **kinetoplastids**, whose mitochondrial DNA is arranged in **kinetoplasts** (**Figure 20-6**).
- **Stramenopiles** are a genetically classified group of diverse protists. They include the filamentous **water molds** (**Figure 20-7**); the **diatoms**, which are phytoplankton with silica shells (**Figure 20-8**); and the **brown algae**, which are single-celled phytoplankton as well as multicellular algae (**Figure 20-9**).
- **Alveolates** have distinct cavities beneath their shells. They include the **dinoflagellates**, which swim using two flagella (**Figures 20-10** and **20-11**); (2) **apicomplexans**, which are nonlocomotive parasites (**Figure 20-12**); and **ciliates**, which are complex unicellular protists that possess **cilia** (**Figures 20-13** and **20-14**).
- **Cercozoans** have elaborate shells and use thin pseudopodia. They include the **foraminiferans**, which have shells made of chalk (**Figure 20-15a**), and the **radiolarians**, which have glassy shells (**Figure 20-15b**).
- **Amoebozoans** move by the use of pseudopodia in both aquatic and terrestrial environments. They include the **amoebas**, which have thick pseudopods and no shells (**Figure 20-16**), and **slime molds**, which are forest-floor decomposers with cellular and acellular forms (**Figures 20-17** and **20-18**).
- **Red algae** are multicellular photosynthetic seaweeds with red pigmentation that masks their chlorophyll (**Figure 20-19**).
- **Green algae** are unicellular, colonial, or multicellular photosynthetic protists, most of which live in ponds and lakes (**Figure 20-20**). They are the closest living relatives of plants.

LEARNING OBJECTIVES

Section 20.1 What Are Protists?

- Describe the three modes of nutrition utilized by protists.
- Discuss the role protists play in the carbon cycle.

Section 20.2 What Are the Major Groups of Protists?

- Discuss the differences in subcellular structures of the various protists.
- Discuss the pathogenic protists and their impact on humans and their crops.
- Discuss the beneficial aspects of protists in the environment.

1. Which of the following correctly describes a protozoan?
   a. prokaryotic
   b. heterotrophic
   c. photosynthetic
   d. cell wall with peptidoglycan

2. Protists acquire food for energy through
   a. ingestion like animals.
   b. absorption like fungi.
   c. photosynthesis like plants.
   d. all of the above.

3. Pseudopods, fingerlike extensions of the plasma membrane, are most often found in protists that
   a. photosynthesize.
   b. ingest their food.
   c. absorb their food.

4. Which of the following structures in a single-celled organisms functions like a stomach in a multicelled organism?
   a. vacuole
   b. pseudopod
   c. eye spot
   d. oral groove

5. Why are vacuoles important for protozoans?
   a. They allow the protist to ingest food that is bigger than the protozoan.
   b. They are used to eliminate wastes from the protozoan.
   c. They stop the enzymes used to digest food from destroying other parts of the protozoan.
   d. None of the above is correct.

6. The large kelp forests that provide food and shelter for marine organisms are
   a. plants.
   b. green algae.
   c. brown algae.
   d. red algae.

7. When hikers drink untreated water from streams, they may suffer severe diarrhea, dehydration, nausea, vomiting, and cramps, caused by Giardia, which is a parasitic
   a. Kinetoplastid.
   b. Euglenid.
   c. Parabasalid.
   d. Diplomonad.

8. The devastating potato famine in Ireland was caused by a water mold from which of the following groups?
   a. Strameophiles
   b. Euglenozoans
   c. Alveolates
   d. Excavates

9. Which of these protists are photosynthetic and bioluminescent?
   a. Euglenids
   b. green algae
   c. water molds
   d. Dinoflagellates

10. Which protists are the most complex single-celled organisms?
    a. Dinoflagellates
    b. Ciliates
    c. Parabasalids
    d. Euglenids

11. Malaria has had a huge impact on the human population. It is caused by a parastic protist called Plasmodium that spends part of its life cycle in mosquitos before moving into humans. Unfortunately, drug-resistant forms of this parasite are spreading across Africa. These parasitic protists are _____.
    a. Apicomplexans
    b. Dinoflagellates
    c. Parabasalids
    d. Diplomonads

12. *Trypanosoma* is a unicellular, eukaryotic blood parasite that causes African sleeping sickness. Into which of the following groups is it classified?
    a. Archaea
    b. Bacteria
    c. Plantae
    d. Protista

13. Which protists are entirely parasitic and have no means of locomotion?
    a. cellular slime molds
    b. Ciliates
    c. Amoebas
    d. Sporozoans
    e. Zooflagellates

14. What phytoplankton group can reproduce so prodigiously that it can cause "red tides," killing large numbers of fish by clogging their gills?
    a. red algae
    b. brown algae
    c. Diatoms
    d. Dinoflagellates

15. Which organisms are sometimes called the "pastures of the sea"?
    a. Dinoflagellates
    b. Foraminiferans
    c. Diatoms
    d. Radiolarians
    e. Amoebae

QUIZ 2

1. The majority of photosynthesis performed on the planet is done by
   a. Cyanobacteria
   b. plants
   c. Protozoans
   d. algae

2. The kingdom of protists is BEST described as
   a. a closely related evolutionary taxon.
   b. eukaryotes that are not plants, animals, or fungi.
   c. the single-celled eukaryotes.
   d. all of the above.
   e. some but not all of the above.

3. Which of the following is NOT a mode of reproduction used by the protists?
   a. asexual reproduction
   b. sexual reproduction
   c. mitosis
   d. binary fission

4. Which of the following is a true statement about endosymbiosis?
   a. The ancestor of chloroplasts is thought to be ancient photosynthetic bacteria.
   b. The ancestor of mitochondria is thought to be ancient aerobic bacteria.
   c. Organelles have multiple membranes resulting from the endocytosis of the bacteria that evolved into the endosymbiotic relationship.
   d. All of the above are correct.
   e. Some but not all of the above are correct.

5. Which of the following organelles evolved first and why?
   a. mitochondria because all living organisms require oxygen
   b. mitochondria because almost all eukaryotes have mitochondria
   c. chloroplasts because all living organisms require energy
   d. chloroplasts because more eukaryotes have chloroplasts than mitochondria

6. Which group of protists has lost its mitochondria?
   a. Strameophiles
   b. Euglenozoans
   c. Alveolates
   d. Excavates

7. Which of these algae dominate in deep, clear tropical waters, where their pigments absorb the deeply penetrating blue-green light and transfer this light energy to chlorophyll, where it is used in photosynthesis?
   a. plants
   b. green algae
   c. brown algae
   d. red algae

8. In the _____, the folds of the inner membrane of the cell's mitochondria have a distinctive shape that appears under the microscope as a stack of discs.
   a. Strameophiles
   b. Euglenozoans
   c. Alveolates
   d. Excavates

9. *Trypanosoma* is a parasitic kinetoplastid that causes a potentially fatal human disease known as
   a. African sleeping sickness.
   b. hiker's sickness.
   c. dysentary.
   d. malaria.

10. Which of the following is NOT a type of green algae?
    a. Volvox
    b. Spyrogyra
    c. Ulva
    d. Kelp

11. Which of the following is an important recycler (decomposer) of nutrients in its ecosystem?
    a. slime molds
    b. sporozoans/apicomplexans
    c. cyanobacteria
    d. viruses
    e. green algae

12. Which protists use flagella for locomotion?
    a. cellular slime molds
    b. Ciliates
    c. Diatoms
    d. Euglenoids
    e. Zooflagellates
    f. (d) and (e)

13. The following tests are performed on an unidentified organism, with the results as shown. How should you classify the organism?
    Chemical A glows when it binds to a plasma membrane. Results = glowing.
    Chemical B turns blue in the presence of chloroplasts. Results = blue color develops.
    Chemical C fizzes when it binds to a nuclear membrane. Results = fizzing.
    Chemical D produces a bad odor when the organism is multicellular. Results = no bad odor.
    a. Cyanobacteria
    b. red algae
    c. Ciliates
    d. Diatoms

14. Which of the following shares a common ancestor with plants and is most like the earliest plants?
    a. Cyanobacteria
    b. slime molds
    c. green algae

d. Amoebas
e. viroids

15. Which of these is a form of phytoplankton that supports aquatic food chains via its photosynthesis?
    a. Sporozoans/Apicomplexans
    b. Diatoms

c. brown algae
d. Amoebas
e. slime molds

ANSWER KEY

Quiz 1

1. b		**9.** d
2. d		**10.** b
3. b		**11.** a
4. a		**12.** d
5. c		**13.** d
6. c		**14.** d
7. d		**15.** c
8. a		

Quiz 2

1. d		**9.** a
2. b		**10.** d
3. d		**11.** a
4. d		**12.** f
5. b		**13.** d
6. d		**14.** c
7. d		**15.** b
8. b		

CHAPTER 21: THE DIVERSITY OF PLANTS

OUTLINE

Section 21.1 What Are the Key Features of Plants?

- Plants have alternating multicellular haploid (**gametophyte**) and diploid (**sporophyte**) generations, with embryos that are dependent on the parent plant for nutrients (**Figure 21-1**).
- Plants are a crucial component of terrestrial ecosystems and provide humans with both necessary resources as well as luxuries.

Section 21.2 What Is the Evolutionary Origin of Plants?

- It is likely that green algae gave rise to the first plants, which were probably similar to modern multicellular green algal forms (**Figure 21-2**).
- Since most green algae live in freshwater habitats, it is likely that their early evolutionary history took place there. The environmental instability of many freshwater habitats may explain how early plants evolved characteristics that allowed them to move successfully to a terrestrial environment.

Section 21.3 How Have Plants Adapted to Life on Land?

- Plant bodies possess a number of features that allow them to resist the effects of gravity and desiccation. These features include (1) the development of roots (or rootlike structures) for water and nutrient absorption from the soil; (2) a waxy **cuticle** and **stomata**, which minimize water loss from the plant; (3) conducting vessels that transport water and nutrients throughout the plant body; and (4) **lignin** molecules that impregnate the plant and stiffen its tissues.
- Plants are well adapted for reproduction in a terrestrial environment. Their adaptive features include **seeds** that provide protection and nutrients to developing embryos, and wind-borne **pollen**, which allows sperm to be easily transferred among plants. Flowering plants also possess **flowers** that attract animals that can act as more effective pollinators, as well as **fruits** that are eaten by animals that then disperse the fruits' seeds.

Section 21.4 What Are the Major Groups of Plants?

- Plants are composed of two major groups: **bryophytes** and **vascular plants** (**Table 21-1**).
- **Bryophytes** include the hornworts, liverworts, and mosses (**Figure 21-3**). Bryophytes are small land plants that lack conducting structures and require water to disperse sperm for reproduction (**Figure 21-4**). Because of this, most bryophytes live in moist environments.
- Vascular plants possess lignin-infused conducting vessels that transport water and nutrients throughout the plant and support its body (allowing the plants to grow larger). Vascular plants consist of two groups: **seedless vascular plants** and **seed plants**.
- **Seedless vascular plants** include the club mosses, horsetails, and ferns (**Figure 21-5** and **21-6**). As with the bryophytes, seedless vascular plant sperm require water for dispersal.
- **Seed plants** produce pollen that allow for wind-borne sperm dispersal. They also produce **seeds** that protect and nourish the developing embryo inside (**Figure 21-7**). Seed plants consist of two groups: **gymnosperms** and **angiosperms**.
- **Gymnosperms** are nonflowering seed plants (**Figure 21-8**). Representative gymnosperm groups include (1) **ginkgos**; (2) **cycads**; (3) **gnetophytes**; and (4) **conifers**, whose seeds develop in cones (**Figure 21-9**).
- **Angiosperms** are flowering seed plants (**Figures 21-10** and **21-11**) and are currently the dominant form of plant. Angiosperms utilize **flowers** to attract pollinators, **fruits** to encourage seed dispersal by animals, and broad **leaves** to capture more light for photosynthesis.

LEARNING OBJECTIVES

Section 21.1 What Are the Key Features of Plants?

- Describe the life cycle of a typical plant.
- Explain the roles of plants in the environment.

Section 21.2 What Is the Evolutionary Origin of Plants?

- Discuss the evolution of plants.

Section 21.3 How Have Plants Adapted to Life on Land?

- Describe the adaptations of plants to allow them to thrive on dry land.

Section 21.4 What Are the Major Groups of Plants?

- Describe the major features of bryophytes and vascular plants.
- Explain the alternating haploid/diploid life cycle of plants.
- Discuss the adaptations of conifers to their habitat.
- Explain the role flowers and fruits play in the flowering plant life cycle.

QUIZ 1

1. Sporophytes produce haploid spores via _____.
   a. meiosis
   b. mitosis
   c. fertilization
   d. pollination

2. The presence or production of _____ distinguishes plants from their nearest relatives, the green algae.
   a. cell walls with cellulose
   b. multicellular, dependent embryos
   c. starch
   d. a nucleus

3. In the life cycle of plants, gametophytes are the
   a. haploid generation, which produces spores.
   b. diploid generation, which produces spores.
   c. haploid generation, which produces gametes.
   d. diploid generation, which produces gametes.

4. Early plant evolution MOST likely occurred in freshwater habitats because _____.
   a. most green algae are freshwater organisms
   b. water temperature can fluctuate seasonally or daily
   c. freshwater habitats may dry up periodically
   d. all of the above

5. Photosynthesis stops during very hot and dry weather because _____.
   a. the stomata close, which cuts off the plant's supply of the carbon dioxide needed for photosynthesis
   b. plants do not make lignin when it is hot and dry
   c. most plants lack vascular tissue that would enable them to absorb water from the soil
   d. the waxy cuticle melts in the hot weather

6. Which of the following is NOT an example of the sporophyte stage of the alternation of generations?
   a. oak tree
   b. moss
   c. pine tree
   d. tomato plant

7. What is the reproductive structure of bryophytes and of seedless vascular plants that encloses eggs and protects them from drying out?
   a. antheridium
   b. archegonium
   c. gametophyte
   d. rhizoids
   e. sporangia

8. The relatively small size of the bryophytes is probably due to _____.
   a. the absence of vascular tissue
   b. the dependence on water for reproduction
   c. their habitat
   d. the lack of true leaves
   e. a unique but conservative pattern of reproduction

9. Even though they are vascular plants, and thus moderately advanced terrestrially, the ferns have not solved all of the problems of terrestrial life. How is this so?
   a. Their zygotes remain unprotected from desiccation.
   b. They are broad leafed.
   c. They do not possess a gametophyte stage.
   d. They lack advanced vascular tissue.
   e. Typically, they lack erect stems.

10. Which of the following is best adapted to a dry habitat?
    a. moss
    b. ferns
    c. horsetails
    d. crabgrass

11. Why has the evolution of reproductive adaptations, the development of pollen and seeds, proven so successful for the gymnosperms and angiosperms?
    a. Both adaptations permit the storage of water for later use.
    b. Both adaptations eliminate the need for dissemination by water.
    c. Both adaptations permit dissemination over long distances through the actions of wind or animals.
    d. (b) and (c) are correct.
    e. All of the above are correct.

12. Which of the following statements about the life cycle of the pine is false?
    a. Pollen grains contain the male gamete.
    b. Pollen grains have wings so they can easily be carried by the wind.
    c. Female gametophyte tissue becomes the food source (nutritive tissue) for the embryo.
    d. Approximately one year after pollination, fertilization occurs, forming the zygote.

13. The anther in a complete flower has the same reproductive function as the _____ in the Bryophyta.
    a. archegonium
    b. sporangium
    c. indusium
    d. antheridium

14. Rust-colored clusters found on the underside of fern fronds are
    a. gametophytes.
    b. sporangia.
    c. archegonia.
    d. antheridia.

15. In the flowering plants, the gametophyte
    a. develops in the flower.
    b. lives independently of the sporophyte.
    c. is the dominant generation.
    d. develops into the plant's fruit.

QUIZ 2

1. All plants produce _____.
    a. spores
    b. seeds
    c. pollen
    d. swimming sperm
    e. fruits

2. Which plant generation is responsible for the production of haploid gametes?
    a. sporophyte
    b. gametophyte
    c. zygote
    d. spore

3. The hypothesis that plants evolved from green algal ancestors is supported by the fact that both plants and green algae
    a. store food as glycogen.
    b. have cell walls made of chitin.
    c. use the same type of chlorophyll and accessory pigments during photosynthesis.
    d. have true roots, stems, and leaves and complex reproductive structures.

4. The rigors of the terrestrial environment led to many adaptations among terrestrial plants. Which of these is NOT a necessary adaptation to dry land?
    a. conducting vessels
    b. cuticle and stomata
    c. lignin
    d. roots or rootlike structures
    e. separate gametophyte stage

5. Which structural adaptation of land plants functions to deliver water and minerals from the roots to the rest of the plant?
    a. cuticle
    b. stomata
    c. conducting vessels
    d. lignin

6. Seed plants produce male gametophytes known as _____.
    a. fruit
    b. flower
    c. pollen
    d. seed
    e. sporangium

7. The sperm of conifers _____.
    a. swim to an egg
    b. are carried in a pollen grain that has tiny wings
    c. are transported to an egg by a bee
    d. are triploid
    e. are found in the ovary of a flower

8. Which major adaptation may be most vulnerable to herbivore attack, especially by insects?
    a. broad leaves
    b. fruits
    c. flowers
    d. roots

9. Flowering plants such as roses and geraniums belong to the group of plants known as _____.
   a. angiosperms
   b. conifers
   c. seedless vascular
   d. bryophytes
   e. gymnosperms

10. Which of the following is TRUE concerning the life cycle of a moss?
    a. Gametophytes are nutritionally dependent upon sporophytes.
    b. Emerging gametophytes arise from germinated spores.
    c. The zygote is protected within specialized tissues of the sporophyte.
    d. Spores are produced within reproductive bodies called antheridia.

11. Which of the following plant groups has swimming sperm and a dominant sporophyte generation?
    a. bryophytes
    b. seedless vascular plants
    c. conifers
    d. angiosperms

12. Which of the following is a characteristic of ferns?
    a. The gametophyte is the dominant generation.
    b. They are vascular plants, containing xylem and phloem.
    c. They produce seeds on frond leaflets.
    d. They produce nonmotile (nonswimming) sperm.

13. The ovule is found in the _____ and _____.
    a. seedless vascular plants, angiosperms or flowering plants
    b. bryophytes, conifers
    c. conifers, seedless vascular plants
    d. conifers, angiosperms or flowering plants

14. The archegonium produces
    a. an egg cell.
    b. a seed.
    c. sperm cells.
    d. a spore cell.

15. The seed contains the embryo of the _____ generation.
    a. gametophyte
    b. sporophyte

ANSWER KEY

Quiz 1

1. a
2. b
3. c
4. d
5. a
6. b
7. b
8. a

9. a
10. d
11. d
12. c
13. b
14. b
15. a

Quiz 2

1. a
2. b
3. c
4. e
5. c
6. c
7. b
8. a

9. a
10. b
11. b
12. b
13. d
14. a
15. b

CHAPTER 22: THE DIVERSITY OF FUNGI

OUTLINE

Section 22.1 What Are the Key Features of Fungi?

- Fungal bodies are composed of filamentous threads (**hyphae**) that form an interwoven mass (**mycelium**). Hyphal threads are ether multicellular or multinucleated (**Figure 22-1**). Fungal cells are surrounded by a cell wall and are typically haploid.
- Fungi secrete enzymes and break down nutrients outside their bodies. They can function as decomposers, parasites, or predators and form mutualistic associations with other organisms.
- Fungi propagate by **spores**, which can be distributed over a wide area (**Figure 22-2**). Fungi can reproduce asexually (by the mitotic production of haploid spores) or sexually (by the meiotic production of haploid spores).

Section 22.2 What Are the Major Groups of Fungi?

- **Chytrid** fungi live in water, which is required for reproduction because they utilize swimming sperm (**Figure 22-4**).
- **Zygomycetes** live in soil or on decaying plant or animal material. Although they can reproduce asexually through the production of haploid spores, they can reproduce sexually by forming diploid spores (**Figure 22-5**).
- **Ascomycetes** reproduce asexually and sexually, but during sexual reproduction, they form a saclike case called an **ascus** (**Figure 22-6**). They include molds and yeasts and can live in decaying forest vegetation (**Figure 22-7**).
- **Basidomycetes** typically reproduce sexually and produce club-shaped reproductive structures called **basidia** (**Figures 22-8** and **Figure 22-9**).

Section 22.3 How Do Fungi Interact with Other Species?

- **Lichens** are symbiotic associations between a fungus and algae (or cyanobacteria, **Figure 22-11**). This arrangement protects algal cells from harsh environments while the fungus receives sugars made by the algal cells (**Figure 22-12**).
- **Michorrhizae** are symbiotic associations between fungi and plant roots (**Figure 22-13**). This arrangement allows plants to absorb more water and minerals from the soil, while the fungi receive sugars from the plant.
- **Endophytes** are symbiotic associations with fungi and plant leaves and stems. This arrangement can protect plant structures from herbivores because of distasteful substances produced by the fungi.
- **Saphrophytes** are fungi that decompose dead organisms, thus acting as recyclers for Earth.

Section 22.4 How Do Fungi Affect Humans?

- Fungi affect humans by attacking useful plants (**Figure 22-14**), but they can attack pest insects and arthropods as well. Fungi also cause human disease (**Figure 22-16**), produce toxins, and produce antibiotic substances (**Figure 22-17**). Fungi are also used to make wine, beer, and bread.

LEARNING OBJECTIVES

Section 22.1 What Are the Key Features of Fungi?

- Describe the various parts of a typical fungi.
- Discuss the differences between plants and fungi in their propagation strategies.

Section 22.2 What Are the Major Groups of Fungi?

- Describe the features of the four phyla of fungi.
- Discuss the benefits of the two types of reproduction in fungi.

Section 22.3 How Do Fungi Interact with Other Species?

- Discuss the different symbiotic relationships between fungi and their hosts.
- Describe the role of fungi in the ecosystem.

Section 22.4 How Do Fungi Affect Humans?

- Discuss the economic impact of fungi on the world food supply.
- Discuss the positive impact on humanity of fungi.

QUIZ 1

1. If you examined the structure of the body of a typical fungus, you would find a mass of
   a. round cells called coccae, which form a long chain of cells connected by chitin.
   b. square cells called hexae, which are clumped together, forming the structures we observe on rotting organic material.
   c. flattened cells called squamae, which are stacked on top of each other, forming a mass called a mycelium.
   d. threadlike cells called hyphae which are only one cell thick.

2. Both fungi and animals _____.
   a. photosynthesize
   b. form embryos
   c. interact to form lichens
   d. fix nitrogen
   e. are heterotrophic

3. The tests that follow have been performed on a cell from an unidentified organism. Based on the results shown, how would you classify this organism?

   Tests:
   Chemical A turns green when a nucleus is present.
   Chemical B bubbles when chloroplasts are present.
   Chemical C pops when chitin is present.

   Results:
   Chemical A turns green.
   Chemical B does not bubble.
   Chemical C pops.
   a. plant
   b. fungus
   c. bacterium
   d. diatom

4. What distinguishes fungal reproduction from that of plants and animals?
   a. No asexual reproduction occurs.
   b. No sexual reproduction occurs.
   c. No embryo is produced when fungi reproduce.

5. Which of the following is associated with the mostly aquatic chytrids?
   a. flagellated spores
   b. cell walls of chitin

c. eukaryotic cells
d. haploid body
e. all of the above.

6. Soft fruit rot and black bread mold belong to which division of fungi?
   a. Ascomycota
   b. Basidiomycota
   c. Deuteromycota
   d. Zygomycota

7. A zygospore undergoes _____ to produce haploid spores.
   a. germination
   b. meiosis
   c. fertilization
   d. mitosis

8. Yeasts, truffles, and Dutch elm disease belong to which division of fungi?
   a. Ascomycota
   b. Basidiomycota
   c. Deuteromycota
   d. Zygomycota

9. The clublike structure producing the spores of typical mushrooms is called _____.
   a. ascus
   b. basidium
   c. conidia
   d. sporangia

10. Which of the following is associated with the group of fungi known as the deuteromycetes?
    a. asexual reproduction
    b. flagellated spores
    c. zygospores
    d. closely related organisms
    e. embryos

11. A test reveals that a lichen contains a prokaryotic symbiont. Which kind of organism is this prokaryotic symbiont?
    a. fungus
    b. heterotrophic bacterium
    c. plant
    d. cyanobacterium
    e. arthropod

12. Which of the following is an important function of fungi?
    a. Yeasts perform alcoholic fermentation that produces CO_2, which makes bread rise.
    b. Some fungi recycle nutrients when they decompose dead plant and animal bodies.
    c. Some fungi produce antibiotics.
    d. All of the above are correct.

13. Which of the following is an example of an economic significance of fungi?
    a. animal disease agents
    b. commercial foods
    c. plant disease agents
    d. symbiosis with plants
    e. all of the above

14. In what ways do some fungi directly affect human health?
    a. as parasites causing disease
    b. as producers of toxins that act as poisons
    c. as a source of antibiotics
    d. all of the above

15. If yeasts are responsible for the alcohol in wine and beer, why don't we get a little tipsy from eating bread?
    a. Alcohol is not produced by yeast inoculated into bread dough.
    b. Fermentation is impaired by the nature of the dough.
    c. Baking the bread evaporates the alcohol produced by the fermenting yeasts.
    d. Yeasts are actually not used in breads, but an ascomycete mold is.

QUIZ 2

1. A haploid asexual spore is formed by a haploid mycelium via _____.
   a. fertilization
   b. meiosis
   c. mitosis
   d. pollination

2. Which of these characteristics is typical of fungi?
   a. autotrophism
   b. possession of cell walls of chitin
   c. cytoplasmic connections between the cells
   d. diploid nature
   e. (b) and (c)

3. The fungal spores resulting from sexual reproduction are produced after
   a. the mycelia of two compatible mating types fuse.
   b. the septa between hyphal cells break down and allow the nuclei to fuse.
   c. a male mycelium encounters a mature female mycelium.
   d. the mycelium encounters a favorable environment for growth.

4. What is the tangled mass of branched filaments that typically forms the fungal body?
   a. conidia
   b. hyphae
   c. mycelia
   d. rhizoids
   e. sporangia

5. One of the more important characteristics used in the classification of fungi is their form of _____.
   a. asexual reproduction
   b. sexual reproduction
   c. mycelium
   d. septa

6. The mushrooms on the pizza you might have eaten last night belong to the phylum
   a. Zygomycota.
   b. Basidiomycota.
   c. Deuteromycota
   d. Ascomycota.

7. The absence of sexual reproductive structures is associated with the
   a. basidiomycetes.
   b. zygomycetes.
   c. deuteromycetes.
   d. ascomycetes.
   e. chytrids.

8. During sexual reproduction in the Basidiomycota and the Ascomycota, _____ produces _____ spores, which give rise to a new mycelium.
   a. meiosis, diploid
   b. meiosis, haploid
   c. mitosis, diploid
   d. mitosis, haploid

9. Which of the following associations is incorrect?
   a. zygomycetes — zygospores
   b. basidiomycetes — fruiting bodies
   c. deuteromycetes — absence of sexual reproduction
   d. chytrids — lightweight, windblown sexual spores
   e. ascomycetes — asci

10. The spores of chytrids are unique because they are
    a. lightweight, windblown structures.
    b. flagellated.
    c. found in club-shaped reproductive structures called basidia.
    d. formed in a saclike case called an ascus.

11. The cells of zygospores contain
    a. a single haploid nucleus.
    b. two haploid nuclei.
    c. a single diploid nucleus.
    d. two diploid nuclei.

12. Which of the following is the most important role fungi have in their ecosystems?
    a. recyclers of nutrients such as carbon and nitrogen from dead animal and plant bodies
    b. parasites on crops such as corn, tobacco, and apples
    c. symbionts with algae or cyanobacteria
    d. pathogens that cause diseases such as ringworm, athlete's foot, and yeast infections
    e. edible fungi such as truffles, morels, and mushrooms

13. What beneficial agricultural role do fungi play?
    a. Fungal pathogens act as fungal pesticides to protect numerous crop species from various insect species.
    b. The application of rusts and smuts alter the coloration of crops confuses potential insect predators.
    c. The American elm and American chestnut have benefited from application of ascomycete fungi.
    d. None of the above is correct.

14. Which of the following is a fungal disease?
    a. athlete's foot
    b. botulism
    c. Lyme disease
    d. malaria

15. Which of these economic problems or diseases is NOT caused by a fungus?
    a. corn smut
    b. Dutch elm disease
    c. histoplasmosis
    d. Mad Cow disease
    e. ringworm

ANSWER KEY

Quiz 1

1. d
2. e
3. b
4. c
5. e
6. d
7. b
8. a

9. b
10. a
11. d
12. d
13. e
14. d
15. c

Quiz 2

1. c
2. e
3. a
4. c
5. b
6. b
7. c
8. b

9. d
10. b
11. c
12. a
13. a
14. a
15. d

CHAPTER 23: ANIMAL DIVERSITY I:
INVERTEBRATES

OUTLINE

Section 23.1 What Are the Key Features of Animals?

- Animals are multicellular, motile heterotrophs that respond to external environmental stimuli and normally undergo sexual reproduction. Their cells lack a cell wall.

Section 23.2 Which Anatomical Features Mark Branch Points on the Animal Evolutionary Tree?

- The anatomical features that mark animal evolutionary branch points are summarized in **Figure 23-1**.
- The earliest branch point is the development of **tissues**, which is expressed in all animals except sponges.
- Animals with tissues can be divided into those that have **radial symmetry** (two embryonic germ layers) and **bilateral symmetry** (three embryonic germ layers, **Figure 23-2**).
- Bilaterally symmetrical animals typically express **cephalization** and some form of body cavity (**Figure 23-3**). Bilaterally symmetrical animals can be divided into two main groups based on patterns of embryonic development: **protostomes** and **duterostomes**.
- **Protosome** animals develop a body cavity within the space between the body wall and digestive cavity. They include the annelids, arthropods, and mollusks.
- **Duterostome** animals develop a body cavity as an outgrowth of the digestive cavity. They include the echinoderms and chordates.

Section 23.3 What Are the Major Animal Phyla?

- Animals of the phylum Porifera (sponges) are aquatic suspension feeders that lack tissues (**Figure 23-4**). Their bodies are generally asymmetrical and are composed of three major cell types (**Figure 23-5**).
- Animals of the phylum Cnidaria (jellyfish, sea anemones, corals, and hydrozoans, **Figure 23-6**) possess tissues, radial symmetry, stinging **cnidocytes** (**Figure 23-8**), a simple nervous system, and a central **gastrovascular cavity** that allows extracellular digestion of food. Most cnidarians alternate between **polyp** and **medusa** forms during their life cycles (**Figure 23-7**).
- Animals of the phylum Platyhelminthes (flatworms) have bilateral symmetry and cephalization but lack a body cavity (**Figure 23-9**). They possess a more advanced nervous system than the cnidarians and can be free living or parasitic (**Figure 23-10**).
- Animals of the phylum Annelida (segmented worms) have a **closed circulatory system**, a well-developed excretory system, and a compartmentalized digestive system. They have a true coelom and **segmented** bodies (**Figures 23-11** and **23-12**).
- Animals of the phylum Mollusca (snails, clams, and squid) are soft bodied and usually protect themselves with some form of shell (**Figure 23-13**). They have gills and annelid-like nervous systems. Most have **open circulatory systems**. Mollusks consist of the **gastropods**, **bivalves**, and **cephalopods** (**Figures 23-14, 23-15**, and **23-16**).
- Animals of the phylum Arthropoda are the most successful animals on earth. They have **exoskeletons** (**Figure 23-17**), jointed appendages, well-developed nervous systems (including **compound eyes**), and specialized respiratory structures that have allowed them to invade nearly every type of aquatic and terrestrial habitat. Some arthropods (the insects) can fly. Arthropods consist of the **insects**, **arachnids**, **myriapods**, and **crustaceans** (**Figures 23-21** to **23-24**).
- Animals of the phylum Nematoda (roundworms) are found nearly everywhere. They lack circulatory and respiratory systems, relying on diffusion for gas and nutrient exchange (**Figure 23-25**). They also possess a pseudocoelom and have free-living and parasitic forms (**Figure 23-26**).
- Animals of the phylum Echinodermata (sea stars, sea urchins, and sea cucumbers) have (bilaterally) symmetrical larvae and radially symmetrical adults (**Figure 23-27**). They possess a nonliving **endoskeleton** that projects through the skin, as well as a **water vascular system** (**Figure 23-28**).

- Animals of the phylum Chordata contain invertebrate and vertebrate groups. Invertebrate chordates are represented by the tunicates and lancelets.

LEARNING OBJECTIVES

Section 23.1 What Are the Key Features of Animals?

- Discuss the basic abilities that define an organism as an animal.

Section 23.2 Which Anatomical Features Mark Branch Points on the Animal Evolutionary Tree?

- Discuss the developmental and anatomical differences between radial symmetry and bilateral symmetry.

Section 23.3 What Are the Major Animal Phyla?

- Discuss the physical attributes and life cycle of sponges.
- Discuss the physical attributes and life cycle of cnidarians.
- Explain the special role of corals in the environment.
- Describe the life cycle of flatworms.
- Describe annelids and their role in the environment.
- Discuss the diversity of mollusks.
- Describe the features of arthropods.
- Explain the benefits provided by an exoskeleton.
- Describe the stages of metamorphosis in invertebrates.
- Discuss the ecological role of insects.
- Discuss the similarities between crustaceans and insects.
- Discuss the impact of roundworms on humans and other animals.

QUIZ 1

1. Which of the following is NOT a characteristic of most animals?
   a. heterotrophism
   b. ability to move
   c. delayed response to external stimuli
   d. sexual reproduction

2. Which of the following characteristics is NOT shared by plants and animals?
   a. multicellularity
   b. eukaryotic nature
   c. sexual reproduction
   d. heterotrophic nature

3. How does radial symmetry differ from bilateral symmetry?
   a. A radially symmetrical animal has dorsal and ventral surfaces. Bilaterally symmetrical animals do not.
   b. Radially symmetrical animals can be divided into symmetrical halves with any plane through the central axis. Bilaterally symmetrical animals can be divided into equal halves by only one specific plane through a central axis.

   c. Radially symmetrical animals possess three embryonic germ layers, while bilaterally symmetrical animals only possess two.
   d. Most radially symmetrical animals are active, free-moving organisms throughout their lives, while bilaterally symmetrical animals are not.

4. An animal with cephalization will have
   a. a cephalic vein.
   b. an anus.
   c. sensory cells/organs and nerve cells clustered at the anterior end of the animal.
   d. a completely lined fluid-filled body cavity.
   e. all of the above.

5. Which of the following is NOT a function associated with a body cavity?
   a. support
   b. digestion
   c. protection of internal organs
   d. allowing internal organs to operate independently of the body wall

6. A true body cavity that develops in the mesoderm is a
   a. coelom.
   b. pseudocoelom.

7. Species of the animal phyla _____ are deuterostomes.
   a. Annelida
   b. Arthropoda
   c. Chordata
   d. Echinodermata
   e. (b), (c), and (d)
   f. (c) and (d)

8. Species of the phyla _____ have pseudocoeloms.
   a. Annelida
   b. Arthropoda
   c. Mollusca
   d. Nematoda
   e. Platyhelminthes

9. A gastrovascular cavity with a single opening is the characteristic digestive system of animals in the _____ phylum.
   a. Arthropoda
   b. Nematoda
   c. Mollusca
   d. Platyhelminthes
   e. Porifera

10. What is the excretory structure in annelid worms?
   a. flame cells
   b. kidney
   c. crop
   d. nephridium

11. Animals in the _____ phyla are segmented.
   a. Annelida
   b. Arthropoda

c. Echinodermata
d. Mollusca
e. (a) and (b)
f. (a) and (d)

12. A radula is a
   a. flexible supportive rod on the dorsal surface of chordates.
   b. stinging cell used by sea anemones to capture prey.
   c. gas-exchange structure in insects.
   d. spiny ribbon of tissue used for feeding in snails.
   e. locomotory structure of sea stars.

13. Which of the following is NOT a mollusk?
   a. barnacle
   b. clam
   c. octopus
   d. slug
   e. snail

14. The success of the cephalopods as predators is supported by the presence of _____.
   a. beaklike jaws
   b. a complex eye
   c. a large and complex brain
   d. tentacles
   e. all of the above

15. The water vascular system unique to echinoderms allows them to _____.
   a. move
   b. do gas exchange
   c. capture food
   d. all of the above

QUIZ 2

1. Animals are distinguished from fungi because animals
   a. are multicellular.
   b. are heterotrophs.
   c. have cells that lack a cell wall.
   d. have cells that are eukaryotic.

2. Which of the following correctly describes animals?
   a. unicellular
   b. heterotrophs
   c. sessile or stationary
   d. prokaryotic

3. How many planes through the central axis will divide an organism with bilateral symmetry into roughly equal halves?
   a. one
   b. two
   c. many

4. Among animals with a fixed body shape, those that are elongated, such as earthworms or scorpions, have _____ symmetry.
   a. no
   b. bilateral
   c. anterior
   d. radial
   e. ventral

5. Which of the following is an example of a coelom?
   a. an air-filled cavity
   b. a fluid-filled cavity
   c. a fluid-filled cavity around the digestive tract that is not surrounded by tissues that are meso-dermal in origin
   d. none of the above

6. Which of the following animal has radial symmetry?
   a. jellyfish
   b. sea star

c. sea anemone
d. sea urchin
e. all of the above

7. All of these animal phyla have coeloms EXCEPT
   a. arthropods.
   b. nematodes.
   c. echinoderms.
   d. mollusks.
   e. annelids.

8. The vast majority of animals
   a. are vertebrates.
   b. lack tissues.
   c. are producers.
   d. lack a backbone.

9. Which of the following is NOT associated with sponges?
   a. epithelial cells
   b. connective tissue
   c. oscula
   d. collar cells
   e. active larvae

10. What are the two functions of the gastrovascular cavity of a cnidarian?
    a. digestion and prey capture
    b. respiration and sexual reproduction
    c. digestion and distribution of nutrients
    d. movement and response to threats
    e. filtering of blood and removal of waste

11. Which of the following is associated with or characteristic of a parasitic flatworm such as a tapeworm?
    a. hooks and suckers
    b. eyespots

c. a respiratory system
d. cilia
e. a complex digestive system

12. Annelids (segmented worms) have many structures comparable to those observed in vertebrates (such as human). Which of the following is a vertebrate structure to which there is nothing comparable in an annelid?
    a. heart
    b. kidney
    c. lung
    d. teeth

13. Animals in the _____ phyla have collar cells.
    a. Porifera
    b. Cnidaria
    c. Annelida
    d. Gastropoda
    e. Chordata

14. Which of the following groups of animals includes the first vertebrates to appear on Earth?
    a. jawless fish
    b. cartilaginous fish
    c. bony fish
    d. cephalopods
    e. none of the above

15. Which of the following animals molts its exoskeleton, allowing the animal to grow larger?
    a. blue crab
    b. bat sea star
    c. scallop
    d. sea urchin
    e. Venus clam

ANSWER KEY

Quiz 1

1. c
2. d
3. b
4. c
5. b
6. a
7. f
8. d

9. d
10. d
11. e
12. d
13. a
14. e
15. d

Quiz 2

1. c
2. b
3. a
4. b
5. d
6. e
7. b
8. d

9. b
10. c
11. a
12. c
13. a
14. a
15. a

CHAPTER 24: ANIMAL DIVERSITY II: VERTEBRATES

OUTLINE

Section 24.1 What Are the Key Features of Chordates?

- Animals of the phylum Chordata have the following features at some point in their development: (1) a notochord; (2) a dorsal, hollow nerve cord; (3) pharyngeal gill slits; and (4) pharyngeal gill slits (**Figure 24-2**).
- Invertebrate chordates are composed of two groups: the **tunicates** and the **lancelets** (**Figure 24-3**).
- Vertebrate chordates have a backbone, which is part of their living endoskeleton.

Section 24.2 What Are the Major Groups of Vertebrates?

- Hagfish (class Myxini) lack jaws and a true backbone; thus, they are not true vertebrates. They reside on the ocean floor and produce copious amounts of slime. Lampreys (class Petromyzontiformes) are jawless vertebrates that often parasitize fish (**Figure 24-4**).
- Jawed fishes can grasp, tear, and crush food. They consist of **cartilaginous fishes**, **ray-finned fish**, and **lobe-finned fish**. **Cartilaginous fish** (sharks, skates, and rays, class Chondrichthyes) are marine predators that lack bony skeletons (**Figure 24-5**). **Ray-finned fish** (class Actinopterygii) are the most diverse vertebrates (**Figure 24-6**) and have webbed fins supported by spines. **Lobe-finned fish** have fleshy fins made of bone and muscle. Some lobe-finned fish (lungfish) have both gills and primitive lungs and are the most closely related to land-dwelling vertebrates (**Figure 24-7**).
- **Amphibians** (class Amphibia) have limbs and simple lungs, allowing for a more terrestrial existence (**Figure 24-8**). Most amphibians need to live near damp environments to keep their permeable skin moist, as well as to facilitate external fertilization and the development of aquatic eggs and larvae.
- **Reptiles** (class Reptilia) have well-developed lungs; scaly, waterproof skin; and shelled, amnioic eggs that cause them to be well adapted for a terrestrial lifestyle (**Figures 24-9** and **24-10**). Reptiles groups include lizards, snakes, crocodilians, turtles, and birds.
- **Birds** are a group of terrestrial reptiles that have endothermic bodies modified for flight (**Figure 24-11**). These modifications include feathers, hollow bones, well-developed respiratory systems, and well-developed circulatory systems.
- **Mammals** (class Mammalia) are endotherms that have hair and well-developed nervous systems and use mammary glands to provide milk to their young. Mammal groups include the monotremes, marsupials, and placental mammals (**Figures 24-13, 24-14,** and **24-15**).

LEARNING OBJECTIVES

Section 24.1 What Are the Key Features of Chordates?

- Describe the four distinctive structures of chordates.

Section 24.2 What Are the Major Groups of Vertebrates?

- Discuss the benefit of developing a true jaw.
- Discuss the differences between cartilaginous and ray-finned fishes.
- Describe the attributes of amphibians.
- Describe the evolutionary adaptations that allowed reptiles to live completely on land.
- Discuss the similarities between birds and reptiles.
- Describe the features unique to mammals.
- Discuss the differences among monotremes, marsupials and placental mammals.

QUIZ 1

1. Pick the chordate characteristic from the following list.
   a. ventral, hollow nerve cord
   b. bony endoskeleton
   c. postanal tail
   d. backbone

2. All chordates share which characteristic?
   a. bilateral symmetry
   b. a fully lined body cavity
   c. a dorsal, hollow nerve cord
   d. pharyngeal gill slits
   e. all of the above

3. In addition to the four characteristics that distinguish chordates from other phyla, chordates also have
   a. ventral, hollow nerve cord
   b. radial symmetry
   c. protostome development
   d. a true body cavity (coelom)
   e. all of the above

4. Which of the following enables you to identify the lamprey species that are parasitic?
   a. suckerlike mouths lined with teeth
   b. complex eyes
   c. fleshy fins
   d. bony skeletons
   e. jaws with rows of razor-sharp teeth

5. Which vertebrate groups is the most diverse but is often overlooked because of humans' habitat bias?
   a. bony fish
   b. jawless fish
   c. mammals
   d. birds

6. The range of amphibian habitats on land is limited by _____.
   a. eggs protected by a jellylike coating
   b. use of their skin as a supplementary respiratory organ
   c. external fertilization
   d. all of the above.

7. Amphibians are most like _____.
   a. mosses
   b. flowering plants
   c. conifers
   d. ferns

8. Reptiles are well adapted to living in drier habitats because of their _____.
   a. hollow bones
   b. production of a shelled amniotic egg
   c. two-chambered heart
   d. moist skin used as a supplemental respiratory organ
   e. external fertilization

9. The ability of birds to fly is facilitated by their _____.
   a. four-chambered heart
   b. lungs supplemented by air sacs
   c. external development in a shelled egg
   d. hollow bones
   e. all of the above.

10. Which of the following characteristics are shared by both arthropods and mammals?
    a. a well-developed nervous system
    b. a closed circulatory system
    c. an internal skeleton
    d. compound eyes

11. The group of terrestrial vertebrates that may be indicators of environmental degradation is _____.
    a. amphibians
    b. bony fishes
    c. lancelets
    d. reptiles

12. The great size and mobility of vertebrates is associated with _____.
    a. four-chamber hearts
    b. lungs used for respiration
    c. lightweight endoskeletons
    d. uterine development of offspring
    e. increased brain size and complexity

13. Which of the following has a ventral nerve cord?
    a. earthworm
    b. shark
    c. coelacanth
    d. frog
    e. hummingbird

14. An animal's ability to live successfully on land is increased by _____.
    a. external fertilization
    b. a two-chamber heart
    c. moist skin used for gas exchange
    d. gills for breathing
    e. development in a shelled egg

15. The high body temperature of birds and mammals is due to _____.
    a. lots of energy lost as heat during metabolism
    b. the presence of sweat, scent, and sebaceous glands
    c. the fur that covers and insulates them
    d. behaviors such as basking in the sun or seeking shade
    e. the exchange of gases and nutrients via the placenta

QUIZ 2

1. Species within animal phyla _____ are deuterostomes.
   a. Annelida
   b. Arthropoda
   c. Chordata
   d. Echinodermata
   e. (b), (c), and (d)
   f. (c) and (d)

2. All members of the phylum Chordata, whether human or lancelet, share certain key features. Which of the following traits is NOT characteristic of all chordates?
   a. dorsal, hollow nerve cord
   b. notochord
   c. pharyngeal gill slits
   d. tail
   e. bony endoskeleton

3. The only chordate feature present in adult humans is the _____.
   a. postanal tail
   b. dorsal, hollow nerve cord
   c. pharyngeal gill slits
   d. notochord

4. Invertebrate chordates differ from vertebrate chordates as a result of the lack of
   a. pharyngeal gill slits.
   b. a postanal tail.
   c. a backbone.
   d. a dorsal, hollow nerve cord.
   e. all of the above.

5. Cartilaginous fish are characterized by _____.
   a. a three-chambered heart
   b. poorly developed lungs
   c. a skeleton formed entirely of cartilage
   d. milk-producing mammary glands

6. What defines, or distinguishes, a mammal from other vertebrates?
   a. its hairless exterior
   b. its primitive, simple brain
   c. milk-producing glands
   d. the fact that most mammals complete the great majority of their development outside the uterus

7. A long period of uterine development and gas, nutrient, and waste exchange between the mother and embryo are characteristic of _____.
   a. all mammals
   b. birds
   c. marsupials
   d. placental mammals
   e. monotremes

8. Phylum Chordata includes all of the following EXCEPT
   a. fish.
   b. birds.
   c. amphibians.
   d. squids.
   e. mammals.

9. Which of the following is a group of invertebrates?
   a. Reptilia
   b. Amphibia
   c. Mammalia
   d. Aves
   e. Echinodermata

10. Class Chondrichthyes includes
    a. whales.
    b. lampreys.
    c. all fish.
    d. frogs.
    e. sharks.

11. Which of these organisms used to be in class Aves but are now placed within the Reptile class?
    a. kangaroo
    b. birds
    c. frogs
    d. dogs

12. Parasitic lampreys have _____.
    a. suckerlike mouths lined with teeth
    b. a backbone
    c. bilateral symmetry
    d. all of the above

13. The vulnerability to both water and air pollutants of which of the following semiterrestrial chordates may be the cause of their dramatic decline in numbers?
    a. amphibians
    b. arthropods
    c. annelids
    d. bony fish
    e. reptiles

14. Reptiles are better adapted to land dwelling than amphibians because of their _____.
    a. two-chamber heart
    b. moist skin used as a respiratory structure
    c. lungs with more surface area for gas exchange
    d. embryo's uterine development
    e. all of the above

15. Reptilian embryos will not dry out in a desert habitat because _____.
    a. reptiles produce lots of defensive slime
    b. reptiles' eggs are protected by a jellylike coating
    c. reptiles' placentas facilitate exchanges between mother and embryo
    d. reptiles produce shelled amniote eggs
    e. reptiles are warm-blooded

ANSWER KEY

Quiz 1

1. c
2. e
3. d
4. a
5. a
6. d
7. a
8. b

9. e
10. a
11. a
12. c
13. a
14. e
15. a

Quiz 2

1. f
2. e
3. b
4. c
5. c
6. c
7. d
8. d

9. e
10. e
11. b
12. d
13. a
14. c
15. d

CHAPTER 25: ANIMAL BEHAVIOR

OUTLINE

Section 25.1 How Do Innate and Learned Behaviors Differ?

- **Innate behaviors** can be performed without prior experience (**Figure 25-1**), whereas **learned behaviors** are modified by experience. Learned behaviors include **habituation**, **trial-and-error learning**, and **insight learning** (**Figures 25-2** and **25-3**).
- Learned behaviors can modify innate behaviors to make them more appropriate for an animal's specific situation (**Figure 25-5**).
- Some forms of learning (e.g., **imprinting**) occur only within innate constraints.

Section 25.2 How Do Animals Communicate?

- An animal **communicates** by producing signals that alter another animal's behavior in a beneficial way for the animal producing the signal.
- **Visual communication** is effective over short distances, whereas **communication by sound** is effective over long distances (**Figures 25-8** to **25-11**). Both methods convey information rapidly.
- **Chemical communication** (such as by **pheromones**) is also effective over long distances, but it persists longer and is difficult to vary in intensity (**Figure 25-12**).
- **Communication by touch** is a method of forming social bonds among members of an animal group and can be used by some animals for sexual communication (**Figure 25-13**).

Section 25.3 How Do Animals Compete for Resources?

- **Aggressive behavior** (e.g., visual and vocal displays or ritualized combat) helps secure resources from members of the same species (**Figures 25-14** and **25-15**). These displays rarely result in serious injury.
- **Dominance hierarchies** are used to manage aggressive interactions, which can disrupt other important tasks (**Figure 25-16**). Dominant animals generally have access to the most resources and/or mates.
- **Territoriality** minimizes aggressive encounters when animals manage and defend boundaries of specific areas that contain resources (**Figure 25-18**).

Section 25.4 How Do Animals Find Mates?

- For reproduction to occur, potential mates must first recognize each other as members of the same species and opposite sex and that they are sexually receptive.
- Finding mates is accomplished by **acoustic** mating signals, **visual** mating signals (**Figures 25-21, 25-22,** and **25-23**), and **chemical** mating signals (**Figure 25-24**).

Section 25.5 What Kinds of Societies Do Animals Form?

- Living in a group has both advantages and disadvantages, with the degree of sociality varying among species.
- Some species, particularly insects and mammals, form complex societies that may sacrifice individuals for the good of the group (**altruism**, **Figure 25-25**).
- Honeybees and naked mole rats have formed particularly complex societies whose members follow specific roles throughout their lives (**Figure 25-27**).

Section 25.6 Can Biology Explain Human Behavior?

- Studies on newborn infants suggest strong correlations between some behaviors and physiology (e.g., suckling and emotional facial expressions, **Figure 25-28**).
- Cross-cultural behavioral comparisons suggest an innate human signaling system, such as demonstrated in facial expressions.
- Studies on identical twins can reveal the genetic basis for human behavior, especially in those who were separated at birth.

LEARNING OBJECTIVES

Section 25.1 How Do Innate and Learned Behaviors Differ?

- Explain the basis of innate behaviors.
- Describe the difference between innate and learned behaviors.
- Discuss the difference between habituation and conditioning.
- Discuss the innate aspect of bird migration.

Section 25.2 How Do Animals Communicate?

- Describe the different types of visual communication.
- Discuss the different types of sounds used to communicate.
- Explain the benefits and constrains of utilizing chemicals to communicate.
- Describe the different organisms utilizing touch for communication.

Section 25.3 How Do Animals Compete for Resources?

- Describe the positive aspects of animal aggression.
- Explain how organisms manage aggression.
- Discuss the role of territory in resource allocation.

Section 25.4 How Do Animals Find Mates?

- Discuss the variety of ways animals obtain mates.

Section 25.5 What Kinds of Societies Do Animals Form?

- Discuss the advantages and disadvantages of social living.
- Discuss altruism as seen in animal societies.
- Discuss the role altruism plays in natural selection.

Section 25.6 Can Biology Explain Human Behavior?

- Describe the innate behaviors seen in humans.
- Discuss the scientific evidence for genetically based personality traits.
- Discuss the beneficial aspects of play.

QUIZ 1

1. Innate behavior is _____.
   a. any observable response to external or internal stimuli
   b. a behavior performed reasonably completely the first time
   c. behavior that is changed on the basis of experience
   d. a change in the speed of random movements
   e. a directed movement toward or away from a stimulus

2. Learned behavior is _____.
   a. any observable response to external or internal stimuli
   b. a behavior performed reasonably completely the first time
   c. behavior that is changed on the basis of experience

   d. a change in the speed of random movements
   e. a directed movement toward or away from a stimulus

3. Which of the following is NOT a type of innate behavior?
   a. movements
   b. fixed action patterns
   c. feeding
   d. habituation

4. The ability to change a behavior as a result of new experiences (i.e., learning) is most closely associated with which of the following behaviors?
   a. instincts
   b. habituation
   c. reflexes
   d. all of the above

5. Fish hatchery workers know that if they want artificially spawned salmon to return to a specific stream to spawn when they reach adulthood, they have to raise the fry (baby salmon) in water from that stream during a certain critical time in the baby salmon's early development. This is a clear demonstration of _____.
   a. operant conditioning
   b. trial-and-error learning
   c. habituation
   d. imprinting
   e. none of the above

6. Bees can perform the dances they use to communicate the location of food sources completely and accurately the first time without prior tutoring or experience. As a result, we call this behavior
   a. innate.
   b. insight learning.
   c. operant conditioning.
   d. trial-and-error learning.

7. Which of the following is an example of one animal's attempt to alter another's behavior?
   a. When threatened, a water moccasin (cottonmouth snake) opens its mouth in a wide gape, exposing the bright white lining of the mouth.
   b. A common flicker (a form of woodpecker) spends significant amounts of time drumming its beak on an aluminum sculpture.
   c. Kangaroo rats living in the desert Southwest rub oils from their skin into the soil at communal sand-bathing areas.
   d. A male spider vibrates a female's web in a specific sequence prior to attempting to mate.
   e. All of the above are correct.

8. Social leaf-cutter ants lay down pheromone trails to lead other members of their colony to a rich food source they have found. This behavior is a classic example of
   a. communication.
   b. habituation.
   c. insight learning.
   d. social hierarchy.

9. In the honeybee waggle dance, the distance of the food source from the hive is represented by the
   a. angle of the straight run relative to the vertical.
   b. duration of the straight run.
   c. intensity of the buzzing during the straight run.
   d. sweetness of food samples given by the dancer to potential recruits.

10. Territoriality _____.
    a. is a behavior in which an animal shares resources with another of its species
    b. increases the aggressive encounters an individual is likely to have
    c. usually leads to a reduction in fitness
    d. involves the active defense of an area containing important resources

11. Trimming the feathers on the left side of a peacock's tail but leaving the right side unaltered would MOST likely _____ the male's chance of mating.
    a. increase
    b. decrease
    c. have no effect on

12. Which of the following is UNLIKELY to be a consequence of sociality?
    a. conservation of energy
    b. ease of finding mates
    c. increased access to limited resources such as food or nest sites
    d. an increase in foraging efficiency
    e. an increased ability to deter predators

13. An individual benefits from a social grouping with other animals through increased
    a. ability to detect, repel, or confuse predators
    b. hunting efficiency
    c. likelihood of finding mates
    d. efficiency resulting from division of labor
    e. all of the above

14. An advantage of using identical twins is that in studies on human behavior
    a. you get twice as much information from twins compared to nontwins.
    b. twins get along well.
    c. genetic influences on behavior can be factored out.
    d. they allow you to factor out environmental influences on behavior.

15. Which of the following is NOT a feature of play?
    a. It occurs more often in young animals than in adults.
    b. It may borrow movements from other behaviors.
    c. It uses considerable energy.
    d. It always has a clear, immediate function.
    e. It is potentially dangerous.

QUIZ 2

1. Grasshopper mice (genus *Onychomys*) are insectivores. Among their prey are beetles that, when threatened, elevate their rear ends and eject a spray of acetic acid into the faces of potential predators. Adult grasshopper mice avoid this by plunging the beetle's rear end into the ground and biting off its head. Juvenile grasshopper mice, when first exposed to these beetles, perform the "jam and bite" behavior but do not always get the right end of the beetle up! Which of the following is the best explanation for this feeding behavior?
   a. The "jam and bite" behavior is innate but modified by learning.
   b. The behavior is a fixed action pattern.
   c. The behavior is completely learned.

2. The rove beetle lays its eggs in the nest of a particular species of ant. Beetle larvae produce a scent that causes the ants to carry the larvae into the ants' brood chamber, where the larvae proceed to eat ant eggs and larvae. Later, the adult beetles mimic the ants themselves in order to get food. An adult beetle approaches an ant and touches the ant with its antennae. Then, it touches its forelegs to the ant's mandibles. This causes the ant to regurgitate sweet, nutritious fluids for the beetle to eat. The adult beetle behavior is an example of the use of _____.
   a. pheromones to evoke a fixed action pattern
   b. pheromones to evoke habituation
   c. touch to evoke classical conditioning
   d. touch to evoke a fixed action pattern

3. Male white-crowned sparrows have distinctive songs that identify not only their species, but also the general region in which they live. In the wild, male birds first begin singing a soft "subsong" at 150 to 200 days after hatching; over time, this develops into the full song characteristic of the species and region in which they live. Experiments with lab-reared birds reveal that males can learn only the song of white-crowned sparrows—they cannot learn even similar songs of closely related species. And, in order for full song to develop properly, males must be exposed to the songs of other white-crowned sparrows between 10 and 50 days after hatching and they must be able to hear themselves sing when they start producing subsongs at 150 to 200 days after hatching (if they are deafened before they begin producing subsong, their full song will not develop). Based on this information, which of the following is correct?
   a. The song is a fixed action pattern.
   b. The song is an innate behavior modified by experience.
   c. Song development is an example of unconstrained learning.
   d. Song development requires imprinting.

4. Which of the following is NOT likely to be a behavior that can be explained in biological terms?
   a. similar behaviors in identical human twins
   b. responses to pheromones among members of the same species
   c. avoidance of mating with closely related individuals
   d. all of the above
   e. none of the above

5. A female bird will continue to feed a parasitic cuckoo chick in its nest even after the chick has killed the bird's own offspring and grown larger than its foster parent. All the cuckoo needs to do is open its mouth and food is inserted. This indicates that feeding of a gaping mouth by an adult bird is _____.
   a. a behavior easily modified by experience
   b. a purely learned behavior
   c. a behavior that has equally large innate and learned components
   d. an innate behavior not easily modified by experience
   e. unknowable based on the information provided.

6. The hybrid offspring produced by mating a member of the eastern European population of blackcaps (a bird) to a member of the western European population of the same species follow a migration route intermediate between that of their parents. This occurs even if the offspring are raised in isolation. From this information, we can reasonably conclude that migration in this species _____.
   a. uses the stars to determine direction
   b. is a purely learned behavior
   c. is controlled entirely by a single gene pair
   d. is imprinted on the birds when they are very young
   e. has a genetic component

7. Which of the following forms of communication would be the MOST likely to be effective over long distances?
   a. communication by sound
   b. communication by touch
   c. communication by visual cues
   d. communication using chemical cues
   e. (b) and (c)
   f. (a) and (d)

8. Insect pests such as gypsy moths and Japanese beetles can be effectively controlled by setting out traps baited with synthetic sex pheromones of the species in question. This type of control is considered better than using chemical pesticides because _____.
   a. the pheromones are cheaper than chemical pesticides
   b. the pheromone traps will kill a broad range of pest species

123

c. the pheromones are very species-specific, so no useful insects are affected

d. the pheromone traps will kill both males and females, while chemical pesticides kill only females

e. all of the above

9. Which of the following types of communication would be LEAST appropriate in the dark confines of a beehive?

a. chemical messages such as pheromones

b. sound signals such as buzzing

c. touch

d. visual signals

10. The basis for viewer interest in the popular television show *Survivor*, as well as its spinoffs, stems from our fascination with which of the following behaviors?

a. territoriality

b. dominance hierarchy

c. aggression

d. all of the above

11. Which of the following is an important function of mating behaviors?

a. communicate species identity

b. communicate gender

c. communicate sexual receptivity

d. defuse aggressive responses

e. all of the above

12. Which of the following conditions would be MOST likely to force animals to associate into loosely organized social groupings?

a. when weather conditions are extreme

b. when prey numbers are low but predator numbers are high

c. when predator numbers decrease but prey numbers remain high

d. (a) and (b)

e. all of the above

13. Small birds will often mob large predatory birds such as owls and hawks. This behavior involves the smaller birds gathering or flying near the larger bird, calling loudly, and even striking the larger bird and may result in the larger bird leaving the area. This behavior is risky for the smaller birds but is commonly seen. From this description, we can reasonably conclude that _____.

a. birds are not intelligent enough to know what is good for them

b. there must be some advantage of this behavior, such as protecting their chicks, that makes the risks worthwhile

c. these birds must have learned this behavior by operant conditioning

d. this behavior is clearly maladaptive and should be eliminated by natural selection

14. Naked mole rats are unique among mammals because

a. they live in colonies

b. most of their lives are spent underground

c. even adults have little body hair

d. most individuals never reproduce

e. they use pheromones to communicate

15. Without thinking about it, some mothers begin lactating when they hear a baby crying. This indicates that lactation in humans is _____.

a. a learned behavior

b. an innate response to a specific stimulus

c. triggered by any vocal stimulation

d. a maladaptive behavior

e. the result of imprinting

ANSWER KEY

Quiz 1

1. b
2. c
3. d
4. b
5. d
6. a
7. e
8. a

9. b
10. d
11. b
12. c
13. e
14. c
15. d

Quiz 2

1. a
2. d
3. d
4. e
5. d
6. e
7. f
8. c

9. d
10. d
11. e
12. d
13. b
14. d
15. b

Chapter 26: Population Growth and Regulation

OUTLINE

Section 26.1 How Does Population Size Change?

- **Populations** grow when births and immigrants exceed deaths and emigrants. Populations decline when the reverse is true.
- The size of a stable population is regulated by (1) the maximum rate a population can increase under ideal conditions (**biotic potential**) and (2) the abiotic and biotic limits on population growth (**environmental resistance**).
- Unchecked, the biotic potential of a population will result in **exponential growth** (**Figure 26-1**). Biotic potential is influenced by (1) an organism's earliest reproductive age, (2) frequency of reproduction, (3) number of offspring produced per reproductive event, (4) an organism's reproductive life span, and (5) death rate under ideal conditions (**Figure 26-2**).

Section 26.2 How Is Population Growth Regulated?

- A population undergoing **exponential growth** will either stabilize or undergo **boom-and-bust cycles** due to the influence of environmental resistance (**Figures 26-3** and **26-4**). Exponential growth occurs most commonly when organisms invade new habitats with abundant resources (**Figure 26-5**), which is commonly observed in **invasive species**.
- Environmental resistance restrains population growth by increasing death rate and decreasing birth rate, resulting in **logistic population growth** (**Figures 26-6** and **26-8**).
- The formula for logistic growth includes a variable for **carrying capacity** (**K, Figure 26-6a**), which defines the maximum sustainable population size for an ecosystem. If K is exceeded, population size may (1) oscillate around K, (2) crash and stabilize at a lower K, or (3) reach zero, resulting in the elimination of that population (**Figure 26-6b**).
- **Density-independent factors**, such as climate and weather, limit populations independently of population density.
- **Density-dependent factors**, such as predation, parasitism, and competition, limit populations more effectively as population density increases (**Figures 26-9** to **26-12**).

Section 26.3 How Are Populations Distributed in Space and Time?

- Populations exhibit three different spatial distributions: (1) **clumped**, (2) **uniform**, and (3) **random** (**Figure 26-13**). Members of a **clumped distribution** population live in groups. Members of a **uniform distribution** population maintain relatively constant distance between individuals. Members of a **random distribution** population do not form social groups and have resources available throughout their habitat.
- Populations show differences in the likelihood of survival (survivorship) at different ages, which can be represented as **survivorship curves** (**Figure 26-14**). **Late-loss survivorship curves** are convex—individuals produce few offspring with low juvenile death rates. **Constant-loss survivorship curves** are relatively constant—individuals have an equal chance of dying at any time during their life span. **Early-loss survivorship curves** are concave—individuals produce large numbers of offspring that have a low chance of survival.

Section 26.4 How Is the Human Population Changing?

- Human population growth rate is currently following a J-shaped exponential model (**Figure 26-15**), primarily due to high birth rates and the **technical, cultural, agricultural**, and **industrial-medical revolutions** that have occurred throughout human history.
- Age structure diagrams show human age groups for males and females in different populations. Expanding populations (as in Mexico) are represented by pyramidal age structure diagrams (**Figure 26-18a**). Stable populations (as in Sweden) are represented by column-shaped age structure diagrams (**Figure 26-18b**). Shrinking populations (as in Italy) are represented by age structure diagrams that are constricted at the base (**Figure 26-18c**).

- Most humans live in developing countries with growing populations (**Figure 26-19**). Although the birthrates of many of these populations have declined, they are still going through **demographic transition**, so their populations continue to grow (**Figure 26-16**).
- The population of the United States is the fastest growing of all developed nations, primarily due to high immigration rates and birthrates (**Figure 26-21**). This has significant environmental implications because the average U.S. citizen uses five times the energy as a citizen of other countries.

LEARNING OBJECTIVES

Section 26.1 How Does Population Size Change?

- Discuss how biotic potential and environmental resistance contribute to the stability of populations.
- Discuss the factors influencing biotic potential.

Section 26.2 How Is Population Growth Regulated?

- Discuss the factors that contribute to a "boom and bust" cycle.
- Describe the different factors leading to exponential population growth.
- Discuss the reasons behind invasive species' exponential growth.
- Discuss the role carrying capacity plays in logistic population growth.
- Describe the different density-independent and density-dependent factors that limit population growth.

Section 26.3 How Are Populations Distributed in Space and Time?

- Describe the three types of population distributions.
- Discuss the three patterns of mortality in populations.

Section 26.4 How Is the Human Population Changing?

- Discuss human population changes over time.
- Describe the factors that allow the continued increase in human population.
- Discuss the differences in human populations between developed countries and developing countries.
- Discuss how the average age of individuals in a country influence future population changes.

QUIZ 1

1. Which of the following would NOT be a suitable research project for an ecologist?
   a. the effect of parasitic worms on death rates of people in tropical Africa
   b. the recovery of the forest community around Mount St. Helens following its eruption in 1980
   c. adaptations of deep-sea fish that allow them to live under such high pressures
   d. how predation by feral cats on ground-nesting birds on an island affects the numbers of each species
   e. how the three-dimensional structure of the active site of an enzyme involved in converting glucose to glycogen in the liver affects its activity

2. Which of the following processes would contribute to an increase in population size?
   a. birth
   b. death
   c. immigration
   d. emigration
   e. (a) and (c)

3. The rate at which a population reproduces and grows under ideal or optimal conditions is known as the _____.
   a. replacement-level fertility
   b. biotic potential
   c. carrying capacity
   d. environmental resistance
   e. survivorship curve

4. The biotic potential of a species depends on the
   a. age at which the organism first reproduces.
   b. chance of survival to the age of reproduction.
   c. frequency with which reproduction occurs.
   d. average number of offspring produced each time.
   e. length of the reproductive life span of the organism.
   f. all of the above.

5. Which of the following examples illustrates the principle of exponential growth?
   a. aphids whose population numbers decrease consistently from year to year
   b. spiders whose population numbers increase one year but decrease in other years
   c. spider mites whose population numbers double every two weeks for the course of the summer
   d. purseweb tarantulas whose population numbers remain essentially unchanged over time

6. Exponential growth requires that
   a. there is no mortality.
   b. there are no density-independent limits.
   c. the birthrate consistently exceeds the death rate.
   d. a species reproduce very quickly.
   e. the species is an exotic invader in an ecosystem.

7. Growth rate implies a change over time. If a city in California has a population of 7500 people and 100 children were born in the city this past year, what is the birthrate for this city?
   a. 0.013
   b. 13%
   c. 75
   d. 760
   e. less than 1%

8. If the elk population were drastically reduced, perhaps due to disease, which prediction could one make about the effects on the wolf population that preys on the elk?
   a. The wolf population will crash, because there are no other food sources for the wolves but elk.
   b. The wolf population will increase because there will be more dead or weakened prey to eat.
   c. The wolf population will be unaffected because it will change its food source to something else.
   d. The wolves will decline as well but will rebound if the deer/elk population increases.
   e. The wolf population will achieve carrying capacity and maintain an S-shaped curve.

9. Carrying capacity is _____.
   a. the total number of organisms of a species that an environment can support.
   b. the population size always reached by a particular species.
   c. a measurement of the total resources in the environment.
   d. a measurement of the total resources in an environment that are used by one population.
   e. the maximum biotic potential of a species assuming no limitation of resources.

10. Which of the following factors is NOT a density-dependent environmental resistor?
   a. weather
   b. competition
   c. predation
   d. parasitism
   e. lack of food

11. Once established, exotic species introduced into a new area often increase dramatically in numbers for quite some time. This increase is believed to occur because _____.
   a. they lack many of the natural enemies (predators, parasites, etc.) that would keep their numbers in check in their native range
   b. exotic species always have very high population growth rates
   c. there are many vacant niches in most ecosystems that exotic species can readily fill
   d. exotic species always have very high carrying capacities
   e. all of the above

12. A clan of spotted hyenas in Ngorongoro Crater (Africa) gobbling down the carcass of a zebra from which they have chased a lone lioness exhibit _____ among themselves.
   a. contest competition
   b. scramble competition
   c. intraspecific competition
   d. interspecific competition
   e. (b) and (c)

13. Although the death rate exceeds the birthrate, the population in North America and the United States in particular continues to expand. What factor might account for this continued increase?
   a. an unstable economy
   b. failed birth control measures
   c. immigration
   d. none of the above

14. Humans have been able to expand the carrying capacity over the course of recorded history
   a. through advances in technology and medicine.
   b. by co-opting the resources of other species.
   c. by exploiting renewable resources faster than they can be replaced and nonrenewable resources that cannot be replaced.
   d. all of the above.

15. Developing countries, such as Pakistan, that have passed only halfway through the demographic transition have _____.
   a. high growth rates as death rates have already dropped but birthrates remain high
   b. high growth rates as birthrates have already risen but have not yet fallen back down
   c. low growth rates as birthrates have already dropped but death rates remain high
   d. low growth rates as death rates have already risen but birthrates remain low
   e. unchanged growth rates as both birthrates and death rates have declined simultaneously

QUIZ 2

1. How does an increasing death rate in a given population affect the J-shaped curve of an exponentially growing population?
   a. The curve remains J-shaped but the rate of growth is slowed, taking longer to reach a given population size.
   b. The curve becomes more vertical in its growth pattern.
   c. The curve is not affected by the death rates, because it is reflecting the overall growth rate.

2. If the number of births in a population is greater than the number of deaths, which of the following is correct?
   a. The population is increasing in size.
   b. The population is decreasing in size.
   c. The value for the rate of growth (r) is positive.
   d. The value for the rate of growth (r) is negative.
   e. (a) and (c) are correct.

3. Assuming that birthrate and death rate for a population were equal, what would happen to population numbers if emigration exceeded immigration?
   a. Population numbers would decrease.
   b. Population numbers would increase.
   c. Population numbers would remain stable.

4. An ecologist studying a population of ostriches records 1000 individuals at the start of a one-year study. Over the course of the year, he records 100 births and 60 deaths. What is the growth rate, r, for this population?
   a. 0.1
   b. 0.04
   c. 0.4
   d. 40
   e. 100

5. A population of pocket gophers has a growth rate (r) of 0.2 per year. If there are initially 100 individuals, how many pocket gophers do you expect at the end of the first and second years?
   a. 120, 144
   b. 120, 140
   c. 20, 4
   d. 120.2, 120.4
   e. 200, 400

6. Growth rate implies a change over time. If a village in Asia has a population of 750 people and 10 children were born in the village this past year, what is the birthrate for this village?
   a. 760
   b. 13%
   c. 75
   d. 0.013
   e. less than 1%

7. Which of the following factors increases populations of organisms?
   a. birth
   b. death
   c. immigration
   d. emigration
   e. (a) and (c)

8. Environmental resistance may limit the size of populations by
   a. increasing both birthrates and death rates.
   b. decreasing both birthrates and death rates.
   c. increasing death rates and/or decreasing birthrates.
   d. decreasing death rates and/or increasing birthrates.

9. As the size of a snowshoe hare population rises, the number of deaths resulting from predation by lynx often also rises because _____.
   a. the number of encounters between predators and prey will increase when there are more prey around
   b. some of the lynx will switch from other prey such as grouse to snowshoe hares as the latter's numbers increase
   c. the increased food available to lynx will eventually increase the numbers of lynx by increasing their birthrate
   d. all of the above

10. Which of the following would NOT decrease the carrying capacity of an ecosystem?
    a. depletion of nonrenewable resources
    b. depletion of renewable resources occurring at a slower rate than the ability of a particular renewable resource to recover
    c. depletion of renewable resources occurring at a rate that exceeds the ability of a particular renewable resource to recover
    d. none of the above

11. The number of individuals of a particular species that the local environment will support is called _____.
    a. biotic limit
    b. trophic level
    c. biotic potential
    d. carrying capacity
    e. ecological maximum

12. Which of the following factors is LEAST likely to influence population size in a density-dependent way?
    a. predation
    b. competition
    c. emigration

129

    d. climate and weather

    e. parasitism and disease

13. Which of the following factors contributed the MOST to the rise in human populations?

    a. the cultural revolution (making tools and improving hunting abilities)

    b. the agricultural revolution (domesticating plants and animals)

    c. industrial-medical revolution (development of mechanical technology and advances in medicine)

    d. information age

    e. biotechnology age

14. Which of the following areas has the highest human population growth rate, with a fertility rate of more than 2.5%?

    a. Africa

    b. China

    c. Central and South America

    d. North America

    e. Asia (excluding China)

15. Why is the U.S. population rate growing as rapidly as it is despite the fact that its fertility rate is 2.03, below the replacement-level fertility of 2.1 children per female?

    a. Even though our fertility rate is below RLF = 2.1, for many years after World War II the fertility rate exceeded the replacement level; therefore, the population is still going up because more women are reproducing.

    b. emigration

    c. immigration

    d. (a) and (c)

ANSWER KEY

Quiz 1

1. e	**10.** a
2. e	**11.** a
3. b	**12.** e
4. f	**13.** c
5. c	**14.** d
6. c	**15.** a
7. a	
8. d	
9. a	

Quiz 2

1. a	**9.** d
2. e	**10.** c
3. a	**11.** d
4. b	**12.** d
5. a	**13.** c
6. d	**14.** a
7. e	**15.** d
8. c	

CHAPTER 27: COMMUNITY INTERACTIONS

OUTLINE

Section 27.1 Why Are Community Interactions Important?

- An ecological **community** is the biotic component of an ecosystem.
- Community interactions limit population size, and, as such, the interactions between community populations act as agents of natural selection (**coevolution**).

Section 27.2 What Is the Relationship Between Ecological Niche and Competition?

- An **ecological niche** describes all aspects of a species' habitat and its interaction with biotic and abiotic environments.
- **Interspecific competition** occurs when the niches of two community populations overlap, resulting in competition for resources. This limits the size and distribution of competing populations. No two species can occupy identical niches continuously (**competitive exclusion principle**, **Figure 27-1**). Species within communities have evolved mechanisms to reduce niche overlap, allowing for **resource partitioning** (**Figure 27-2**).
- **Intraspecific competition** is the most intense form of competition because individuals of the same species have identical ecological niches.

Section 27.3 What Are the Results of Interactions Between Predators and Their Prey?

- Predators and prey act as mutually powerful agents of natural selection.
- Predators have evolved traits (e.g., **camouflage**, counteracting behaviors, **aggressive mimicry**, and venom) that allow them to improve their ability to catch prey.
- Prey animals have evolved traits (e.g., **camouflage**, **warning coloration**, **mimicry**, and poisons) that allow them to improve their ability to avoid predation.

Section 27.4 What Is Symbiosis?

- **Symbiotic relationships** describe the close interaction between organisms of different species over an extended time. These relationships include **parasitism**, **commensalism**, and **mutualism**.
- **Parasitism** occurs when a parasite feeds on a host, often harming it but not usually killing it—for example, a tapeworm living within a human intestine.
- **Commensalism** occurs when one species benefits, but the other is unaffected by the relationship—for example, barnacles living on a whale.
- **Mutualism** occurs when both species benefit from their relationship (**Figure 27-13**)—for example, algae and a fungus as components of a lichen.

Section 27.5 How Do Keystone Species Influence Community Structure?

- **Keystone species** have a disproportionate influence on community structure. Removal of a keystone species can dramatically alter community structure (**Figure 27-14**).

Section 27.6 Succession: How Do Community Interactions Cause Change over Time?

- **Succession** describes the progressive change over time in the types of populations that form a community (**Figure 27-15**). There are two forms of succession: **primary succession** and **secondary succession**.
- **Primary succession** can take thousands of years and occurs in areas where no community existed previously (**Figure 27-16**).
- **Secondary succession** occurs more rapidly (hundreds of years) as it builds upon a disrupted, previously existing community (**Figure 27-17**).
- Uninterrupted succession results in a **climax community** that is stable and self-perpetuating.

LEARNING OBJECTIVES

Section 27.1 Why Are Community Interactions Important?

- Discuss the role of community interactions on natural selection.

Section 27.2 What Is the Relationship Between Ecological Niche and Competition?

- Explain the concept of ecological niche.
- Discuss interspecific competition in relation to a species' niche.

Section 27.3 What Are the Results of Interactions Between Predators and Their Prey?

- Discuss coevolution between predator and prey.
- Describe how mimicry is used by both predators and prey.
- Discuss the use of chemicals in the predator/prey relationship.

Section 27.4 What Is Symbiosis?

- Discuss the differences among commensal, parasitic, and mutualistic relationships between organisms.

Section 27.5 How Do Keystone Species Influence Community Structure?

- Describe the role of a keystone species.

Section 27.6 Succession: How Do Community Interactions Cause Change over Time?

- Discuss the differences between primary and secondary succession.
- Describe the characteristics of a climax community.
- Discuss the benefits provided by maintaining a subclimax community.

QUIZ 1

1. When groups of species that interact with each other in a given area are considered together, the level of organization is called the _____.
   a. biome
   b. biosphere
   c. population
   d. community
   e. ecosystem

2. Following applications of insecticides to agricultural fields to control pest insects, sometimes the crops suffer *more* insect damage than if no pesticides were applied. Knowing what you know about interactions between species and the factors that regulate population sizes, what is the most likely explanation for this?
   a. The insecticides probably killed off most of the predators that previously kept the size of the plant-eating insect population from exploding. The insects then increased in number and did more damage to the plants.
   b. The insecticides probably killed off most of the competitors that previously kept the size of the plant-eating insect population from exploding.

   The insects then increased in number and did more damage to the plants.
   c. The insecticides probably weakened the defenses of the plants, making them more vulnerable to attack by the surviving plant-eating insects.
   d. The insecticides probably inhibited the growth of mycorrhizae and nitrogen-fixing bacteria on which the plants rely for nutrients. The plants were therefore weakened and more prone to attack by insects.

3. No two species can occupy the same ecological niche in the same place at the same time. This statement is known as _____.
   a. mutualism
   b. a premating isolating mechanism
   c. Darwin's theory of natural selection
   d. the Hardy-Weinberg principle
   e. the competitive exclusion principle

4. If the island population of fruit bats is competing with other nocturnal mammals, such as the sugar glider, for the same resources of fruit, pollen, nectar, and insects, which concept best describes the splitting of this similar niche of food resource so

that both species may coexist with these limited resources?

   a. resource partitioning
   b. competitive exclusion principle
   c. adaptation
   d. mimicry
   e. mutualism

5. A habitat is best defined as
   a. the home or location where an organism lives.
   b. the occupation of the organism in the ecosystem in which it resides.
   c. all the organisms and their nonliving environment within a defined area.
   d. all the interacting populations within an ecosystem.

6. What is the function of aggressive mimicry?
   a. to hide a prey from a predator
   b. to warn a predator that a prey is dangerous
   c. to warn a predator that a prey is distasteful
   d. to keep prey from recognizing a predator
   e. to startle a prey when it sees a predator

7. The rattles of rattlesnakes would be similar in function to which of the following?
   a. camouflage
   b. warning coloration
   c. startle coloration
   d. aggressive mimicry
   e. (b) and (c)

8. The close interaction between organisms of different species over an extended period of time in which one individual benefits while the other individual neither benefits nor is harmed by the relationship is known as _____.
   a. predation
   b. competition
   c. parasitism
   d. mutualism
   e. commensalism

9. Honeyguides are African birds that excitedly lead the way to a bee's nest, and ratels are the honey- and bee-eating mammals that open up and scatter the contents of the bee's nests, allowing both the ratels and the honeyguides to feed on the contents. This relationship is an example of _____.
   a. predation
   b. competition
   c. parasitism
   d. mutualism
   e. commensalism

10. *Trypanosoma* is a protozoan (single-celled organism) that lives and reproduces for an extended period in the blood of a mammalian host (e.g., a human, native antelope, or introduced cattle). Newly introduced cattle generally die from this infection if they are not treated, whereas the native antelope or cattle that have been exposed to this protozoan for several generations are less severely affected. This relationship is an example of _____.
   a. predation
   b. competition
   c. parasitism
   d. mutualism
   e. commensalism

11. The beaver, *Castor canadensis*, seldom reaches densities anywhere near those achieved by most other rodents. However, its effect on ecosystems can be enormous. By damming streams, it creates ponds and wetlands. This can kill trees in the midst of an otherwise continuous stand of climax forest, making room for many aquatic and semiaquatic species. When the beaver move on and the pond fills in, the meadow left behind also represents a new and different habitat type suitable for yet other species. In short, the beaver can dramatically increase the complexity of ecosystems, leaving behind a mosaic of habitat types that supports enormous biodiversity. It is clear from this description that the beaver is a _____ in the forest ecosystem.
   a. keystone species
   b. pioneer species
   c. climax species
   d. dominant species

12. In the intertidal zone, diverse assemblages of many invertebrate species and algae exist attached to the rocks. If an oil spill occurred that directly affected only a species of starfish that is a keystone predator in this system, totally eliminating it, what do you predict would happen to the community?
   a. No significant change in the structure of the community would be likely to occur.
   b. The community is likely to become less diverse, increasingly dominated by a few species that are good competitors for space.
   c. The community is likely to become more diverse, as strong and weak competitors can then coexist.

13. Why does a climax community tend to persist in an area?
   a. It is made up of large, hard-to-move organisms.
   b. The species in it do not alter their environment significantly.
   c. It is able to change the weather.
   d. The species in it are able to kill off competitors.

14. Boulders are often overturned by wave action during storms along the lower reaches of the rocky intertidal shore. Along the coast of Southern California, there is a fairly predictable sequence of species that colonizes the bare rock as spores and then replaces each other. This starts with the green

algae *Ulva*, which is replaced after a year or so by various red algae and so on. This is an example of
_____.

a. primary succession
b. secondary succession
c. aggressive mimicry
d. a keystone species

15. The structural change in a community and the abiotic environment over time, in which assemblages of plants and animals replace one another in a relatively predictable sequence, is called _____.

a. evolution
b. radiation
c. selection
d. succession
e. symbiosis

QUIZ 2

1. Why are community interactions so important?
   a. Interspecies interactions affect the ability of species to survive and reproduce.
   b. Coevolution has shaped the behaviors and form of interacting species.
   c. Both of the above are correct.
   d. Neither of the above is correct.

2. Corals are polyps of coelenterates (cnidarians) that contain numerous algae in their tissues. The algae contain photopigments that give the corals their color. When stressed by high temperatures, water pollution, or similar shocks, many corals expel their algae and turn white, a process called "coral bleaching." This appears to help the coelenterates survive the initial shock, but if they do not recover their native algae quickly, they soon die. Similarly, the algae cannot live for long outside the coelenterates' bodies. Based on this information, the relationship between the two organisms is most likely _____.

   a. a mutualism
   b. interspecific competition
   c. commensalism
   d. parasitism
   e. predation

3. In a farm pond location, a landowner decided to introduce bluegill *Lepomis macrochirus* because she liked to fish and preferred to eat the bluegill species. The species of fish already living in the pond were sunfish *Lepomis humilis*. After introduction of the bluegill, which is very similar to sunfish in habitat and food preferences, the landowner discovered several years later that there were not as many sunfish in the pond; the sunfish numbers had diminished over time. This is an example of
   a. intraspecific competition.
   b. interspecific competition.
   c. competitive exclusion principle.
   d. (a) and (c).
   e. (b) and (c).

4. What were the differences in the niches of the paramecium in Gause's second experiment that allowed them to coexist?
   a. food eaten
   b. body size
   c. feeding area
   d. preferred water temperature
   e. preferred pH

5. Within a year of the abandonment of agriculture on a plot of prairie, the previously bare soil is overrun with annual weeds. Light, carbon dioxide, and mineral nutrients are readily available, but soil moisture is limiting. Still, some species appear to coexist very close to one another. Upon closer examination, two such species, smartweed and bristly foxglove, are observed to have very different root systems and ways of managing water. Smartweed has a deep taproot, extending about a meter beneath the surface, tapping (literally) into a continuous deep water supply. Bristly foxtail has a much shallower and spreading fibrous root system, reaching less than 20 cm down. However, the latter plant is able to tolerate periods of drought and to take up water rapidly after a rain. This example is a clear case of _____.
   a. resource partitioning
   b. competitive exclusion
   c. intraspecific competition
   d. commensalism
   e. a keystone species

6. A poisonous frog with bright and colorful body markings is an example of _____.
   a. camouflage
   b. mimicry
   c. warning coloration
   d. aggressive mimicry

7. Fireflies use their amazing bioluminescence to find mates of the right species. Males fly around and emit a series of flashes in a pattern that differs from all other species. A female, perched nearby,

may respond with her own species-specific signal. The male immediately flies towards her and, after a few more exchanges, it's wedded bliss! Well, not always. Females of a few species have learned to imitate other species' signals and, when their males arrive, they eat them. These *femme fatales* are exhibiting a form of _____.

a. aggressive mimicry
b. startle coloration
c. camouflage
d. warning coloration
e. chemical warfare

8. In many ways, parasitism and predation are similar types of interactions between species. Which of the following is NOT true about their differences?

a. Parasites are usually much smaller than their hosts, but predators are usually larger than their prey.
b. Parasites are usually much more numerous than their hosts, but predators are usually less numerous than their prey.
c. Parasites usually do not kill their hosts immediately, but predators usually do kill their prey immediately.
d. Parasites usually have no effect on their hosts, but predators usually harm their prey.

9. On many coral reefs in the Pacific, large fish bearing parasites will visit "cleaning stations" where species of small fish, known as *cleaner wrasses,* remove parasites and loose scales from the larger fish. The cleaner fish may even enter the mouth and gills of the larger fish to clean parasites from the soft tissues. The cleaner fish are recognized by their coloration, black with a bright blue or yellow stripe, and by a little "dance" that invites the larger fish in to be cleaned. Also on the reef, however, is a small fish known as the *saber-tooth blenny.* Looking and acting very much like the cleaner wrasse, the blenny also attracts the larger fish, but instead of cleaning away parasites, the blenny bites small bits of flesh from the larger fish. The interactions among the larger fish, the cleaner wrasse, and the saber-tooth blenny represent all of the following EXCEPT?

a. mutualism.
b. parasitism.
c. aggressive mimicry.
d. warning coloration.

10. What is the term for a situation in which one organism benefits from its close association with a second species, but the second species is harmed in the process?

a. commensalism
b. mutualism

c. parasitism
d. (a) and (b).

11. In the rocky intertidal zone, along the coast of the state of Washington, space on the substrate (rocks) is a critical resource for sessile (attached) organisms, such as algae, barnacles, and mussels, and also for other organisms that graze the algae or prey on the sessile animals. For example, limpets graze the algae, and three species of barnacle dominate the algae for space. Among them they form a dominance hierarchy, with the larger *Semibalanus*-dominating *Balanus,* which in turn dominates the smaller *Chthamalus.* The mussel, *Mytilus,* is at the top of this hierarchy. A large thaid snail feeds on *Balanus* and *Chthamalus* but cannot take the larger *Semibalanus* or the mussel. The starfish, *Pisaster,* however, prefers to feed on mussels and the large barnacle *Semibalanus.* Thus, if the starfish is present in this community, the community will be a diverse assemblage of nearly all of the species described here; but if the starfish is absent, the dominant mussel will take over most of the available space. In this example we see _____.

a. predation
b. intraspecific competition
c. interspecific competition
d. a keystone species
e. all of the above

12. A keystone species is

a. any species that is found in the state of Pennsylvania.
b. a species that plays a key role in determining community structure that is greater than its abundance alone would predict
c. a type of symbiosis between two species in which one species lives inside another
d. any species that plays an important role in the community
e. none of the above

13. Plants and animals invade a region recently scoured clean by a retreating glacier and over time are replaced by other species. This statement describes _____.

a. a climax community
b. primary succession
c. secondary succession
d. all of the above

14. When all vegetation is removed from a site by human activities or by natural forces such as volcanoes or glaciers, _____ species are the first to colonize the sites.

   a. pioneer
   b. Archaea
   c. climax
   d. latent

15. Organisms in a succession community help cause the changes that result in their own replacement because

   a. they become old and unfit for the environment.
   b. they emigrate to other areas to make room for new species.
   c. they change the physical environment in ways that favor competitors.

ANSWER KEY

Quiz 1

1. d		**9.** d	
2. a		**10.** c	
3. e		**11.** a	
4. a		**12.** b	
5. a		**13.** b	
6. d		**14.** a	
7. e		**15.** d	
8. e			

Quiz 2

1. d		**9.** d.	
2. a		**10.** c	
3. e		**11.** e	
4. c		**12.** b	
5. a		**13.** b	
6. c		**14.** a	
7. a		**15.** c	
8. d			

Chapter 28: How Do Ecosystems Work?

OUTLINE

Section 28.1 What Are the Pathways of Energy and Nutrients?

- Energy moves through ecosystems in a continuous one-way flow, originating as sunlight. Nutrients are constantly recycled within and among ecosystems (**Figure 28-1**).

Section 28.2 How Does Energy Flow Through Communities?

- Energy enters communities through photosynthesis (**primary productivity, Figure 28-2**). The energy that photosynthetic organisms store and make available to other members of the community is called **net primary productivity** (**Figure 28-3**).
- Energy is passed within communities from one **trophic level** to another as a series of feeding relationships. **Autotrophs** (**producers**) capture energy from sunlight and form the first trophic level. **Herbivores** (**primary consumers**) feed on producers and form the second trophic level. **Carnivores** (**secondary consumers**) feed on primary consumers and form the third trophic level. Sometimes carnivores eat other carnivores (**tertiary consumers**) and form the fourth trophic level. **Omnivores** consume plants and animals and occupy multiple trophic levels.
- A linear representation of feeding relationships is called a **food chain** (**Figure 28-4**). Typically, food chains interconnect, forming **food webs** (**Figure 28-5**).
- **Detritus feeders** and **decomposers** digest dead bodies and decaying organic matter. They release the energy stored within them as well as recycle nutrients back into ecosystems.
- Only about 10% of available energy is transferred from one trophic level to the one above it, which can be represented as an **energy pyramid** (**Figures 28-6** and **28-7**).

Section 28.3 How Do Nutrients Move Within and Among Ecosystems?

- A **nutrient cycle** depicts the movement of a particular nutrient from its **reservoir** through the biotic portion of its ecosystem and back to its reservoir.
- In the **carbon cycle**, atmospheric CO_2 enters producers through photosynthesis, then through the food web, and is released back into the atmosphere as a result of cell respiration (**Figure 28-8**).
- In the **nitrogen cycle**, atmospheric nitrogen gas is converted by bacteria (and human industrial activity) into ammonia and nitrate, which are used by plants. This nitrogen enters the food web and reenters the environment by excretion and the activity of decomposers and detritus feeders (**Figure 28-9**).
- In the **phosphorous cycle**, phosphate in rocks dissolves in rainwater and is absorbed by producers and then passes through food webs. Phosphate returns to the soil and water by excretion and the action of decomposers. Some phosphate is carried to the oceans, where it is deposited in sediments (**Figure 28-10**).
- In the **hydrologic cycle**, water in the oceans evaporates and enters the atmosphere. This water returns to Earth as precipitation, which flows into lakes, underground reservoirs, and rivers, which flow into the oceans. Water is absorbed by plants and animals and enters the food webs.

Section 28.4 What Causes "Acid Rain"?

- Human consumption of fossil fuels and industrial processes cause the release of sulfur dioxide and nitrogen oxides into the atmosphere. These compounds are converted to sulfuric acid, which falls to Earth as acid deposition (including acid rain). Acidification of lake and forest ecosystems has reduced their ability to sustain life.

Section 28.5 What Causes Global Warming?

- Elevated atmospheric CO_2, a **greenhouse gas** produced by the burning of fossil fuels, is correlated with increased global temperatures (**global warming, Figures 28-15** and **28-16**).
- Global warming is causing ancient ice to melt (**Figure 28-18**) is predicted to cause extreme weather conditions, and is influencing seasonal wildlife activity.

LEARNING OBJECTIVES

Section 28.1 What Are the Pathways of Energy and Nutrients?

- Discuss energy flow through a community.

Section 28.2 How Does Energy Flow Through Communities?

- Explain the difference between autotrophs and heterotrophs.
- Discuss the organisms typical of each trophic level from producers through tertiary consumers.
- Describe the interconnected relationships found in a food web.
- Discuss the roles of detritus feeders and decomposers in the environment.
- Explain the relationship between trophic level and energy available.

Section 28.3 How Do Nutrients Move Within and Among Ecosystems?

- Discuss the location of carbon reservoirs in the environment.
- Describe the cycling of carbon between reservoirs.
- Explain the process of nitrogen fixation and its importance to life.
- Discuss the unique attributes of the phosphorous cycle.
- Explain the water cycle.
- Discuss the socioeconomic issues surrounding access to fresh water.

Section 28.4 What Causes "Acid Rain"?

- Discuss the causes and effects of acid rain.

Section 28.5 What Causes Global Warming?

- Explain how heat is trapped in the atmosphere.
- Discuss the causes and effects of global warming.

QUIZ 1

1. An ecologist studying a plot of ground in the tundra excludes all herbivores from her study area and estimates the plant biomass at 530 grams per square meter. She comes back to the same plot one year later and estimates that the biomass has increased to 670 grams per square meter. The difference in these two values, 140 grams, represents the _____ for that year.
   a. biological magnification
   b. food chain length
   c. food web complexity
   d. net primary productivity
   e. trophic transfer efficiency

2. Of the solar energy that strikes the outer reaches of Earth's atmosphere, approximately what percentage ends up in carbohydrate molecules produced by photosynthetic organisms?
   a. 0.03%
   b. 1%
   c. 3%
   d. 10%
   e. 90%

3. Earthworms are _____.
   a. detritus feeders
   b. herbivores
   c. primary consumers
   d. producers
   e. secondary consumers

4. The base of the energy pyramid represents _____.
   a. producers
   b. decomposers
   c. primary consumers
   d. secondary consumers
   e. tertiary consumers

5. Which of the following are heterotrophic?
   a. producers
   b. decomposers
   c. primary consumers
   d. secondary consumers
   e. (b), (c), and (d)

6. Of the following trophic levels, which would support the fewest organisms?
   a. producer
   b. decomposer
   c. primary consumer
   d. secondary consumer
   e. tertiary consumer

7. Which of the following trophic levels is always the final link in the food chain?
   a. producer
   b. decomposer
   c. primary consumer
   d. secondary consumer
   e. tertiary consumer

8. Which of the following nutrients remains chemically the same as it is cycled through the food chain and is generally not used in the synthesis of new molecules?
   a. water
   b. carbon
   c. nitrogen
   d. phosphorus
   e. (c) and (d)

9. In the carbon cycle, carbon is returned to the atmosphere by _____.
   a. photosynthesis
   b. evaporation of water
   c. burning of fossil fuels
   d. respiration of plants and animals
   e. (c) and (d)

10. For which of the following nutrients is rock a major reservoir?
    a. water
    b. oxygen
    c. carbon
    d. nitrogen
    e. phosphorus

11. How is nitrogen released back to the atmosphere once it has been incorporated into the body of an organism?
    a. nitrogen fixation
    b. through symbiotic association with a legume

c. by decomposers and denitrifying bacteria
d. (a) and (b)

12. Which of the following is a major contributor to the problem of acid deposition?
    a. oxygen
    b. carbon dioxide
    c. sulfur dioxide
    d. nitrogen oxides
    e. (c) and (d)

13. So far, global warming has been documented to be causing _____.
    a. changes in precipitation patterns on land, with some areas subjected to more severe and frequent droughts, and other areas more frequent and severe floods
    b. melting of ice sheets and retreat of glaciers at unprecedented rates
    c. shifts in the distribution and abundance of a number of plant and animal species
    d. shifts in the timing of spring events, which are occurring much earlier than previously
    e. all of the above
    f. none of the above; to date, global warming has had no effects on any of these events.

14. How does using wood derived from trees as a source of fuel for cooking affect the carbon cycle?
    a. It decreases the uptake of CO_2 from the atmosphere as fewer trees carry on photosynthesis.
    b. It has no effect on the carbon cycle.
    c. It increases the release of CO_2 into the atmosphere as the wood is burned.
    d. (A) and (c) are correct.

15. Increased levels of carbon dioxide and other greenhouse gases in the atmosphere contribute to global warming by
    a. allowing more sunlight to reach Earth.
    b. preventing more of the heat radiated from Earth's surface from escaping into space.
    c. promoting and sustaining forest fires and other forms of combustion that release heat.

QUIZ 2

1. DDT and other substances that undergo biological magnification are dangerous because they are

   _____.
   a. biodegradable
   b. not biodegradable
   c. fat soluble
   d. water soluble
   e. (b) and (c)

2. A spider that feeds on an aphid that in turn feeds on germinating blades of wheat would belong to which consumer category?
   a. herbivore
   b. primary consumer
   c. secondary consumer
   d. tertiary consumer

3. Which of the following animals are omnivores?
   a. deer
   b. wolves
   c. hyenas
   d. black bears
   e. fungi

4. What are the primary consumers of the ocean environment?
   a. phytoplankton
   b. zooplankton
   c. small fish such as anchovies
   d. big fish such as tuna
   e. jellyfish

5. Which of the following organisms is considered to be a detritus feeder or decomposer?
   a. bacteria
   b. worms
   c. fungi
   d. all of the above

6. What is the base of the food pyramid or the food chain in a marine ecosystem?
   a. plankton
   b. zooplankton
   c. phytoplankton
   d. small fish
   e. coral

7. Organisms that eat dead or decaying material to gain their source of energy are called _____ or _____.
   a. autotrophs, consumers
   b. consumers, producers
   c. detritivores, herbivores
   d. detritus feeders, decomposers

8. As one moves from the equator toward the poles, the net primary productivity of forest ecosystems generally _____.
   a. decreases
   b. increases
   c. remains the same
   d. first decreases and then increases
   e. first increases and then decreases

9. Which trophic level has members that require uptake of ammonia, a complex form of nitrogen?
   a. producers
   b. herbivores
   c. carnivores
   d. decomposers

10. In ecosystems, elements such as carbon and nitrogen
    a. are neither created nor destroyed, but may change molecular form as they pass from organism to organism and between abiotic and biotic components.

    b. are produced by the sun, travel to Earth, and pass briefly through ecosystems, being degraded in the process, and are ultimately lost to space.
    c. play no role whatsoever.

11. Which of the following materials cycle through an ecosystem: (1) carbon, (2) nitrogen, (3) oxygen, (4) water, and (5) energy?
    a. (1) through (5)
    b. (1) through (4)
    c. (1) through (3)
    d. only (1) and (2)
    e. only (4) and (5)

12. Why is acid rain, or acid deposition, considered harmful?
    a. Moisture in the air becomes acidified and then falls on plants and the soil below, harming them.
    b. Acid rain leeches essential nutrients out of the soil (e.g., potassium and calcium) and kills decomposers in the soil.
    c. Dead, or weakened, plants make the soil much more susceptible to erosion.
    d. All of the above are correct.

13. Which of the following greenhouse gases does phytoplankton in the marine environment require for the process of photosynthesis, thereby assisting in the regulation of Earth's atmospheric composition?
    a. oxygen
    b. hydrogen
    c. carbon monoxide
    d. carbon dioxide
    e. sulfur dioxide

14. Which of the following is NOT considered a greenhouse gas?
    a. oxygen
    b. water vapor
    c. methane
    d. carbon dioxide

15. Conserving electricity by using fluorescent lights instead of incandescent lights and turning off lights and appliances when not in use _____.
    a. actually increases greenhouse gas production as the predominant fossil fuel–powered electrical generating plants consume large quantities of carbon dioxide when they are operating
    b. has no effect on greenhouse gas production because nearly all electricity in this country is produced by nuclear reactors and renewable wind- and solar-generating stations
    c. helps reduce greenhouse gas production because the majority of electricity in this country is still produced by fossil fuel–powered generating plants. Since these plants produce carbon dioxide when they operate, conserving electricity reduces greenhouse gas production.

ANSWER KEY

Quiz 1

1. d
2. a
3. a
4. a
5. e
6. e
7. b
8. a

9. e
10. e
11. c
12. e
13. e
14. d
15. b

Quiz 2

1. e
2. c
3. d
4. b
5. d
6. c
7. d
8. a

9. a
10. a
11. b
12. d
13. d
14. a
15. c

CHAPTER 29: EARTH'S DIVERSE ECOSYSTEMS

OUTLINE

Section 29.1 What Factors Influence Earth's Climate?

- The climate for a region is determined by the amount of sunlight and water and by the range of temperatures.
- The curvature and tilt of Earth influences climate because they affect the amount of light that strikes different latitudes at different times of the year (**Figure 29-1**).
- Uneven heating of Earth's surface generates **air currents** that produce broad climactic regions (**Figure 29-2**). Earth's rotation, winds, and solar heating of the oceans generates circular patterns of ocean currents (**gyres**) that moderate nearshore climates (**Figure 29-3**).
- As elevations increase, temperatures decrease, creating biomes like those in more extreme latitudes (**Figure 29-4**). As air moves down a mountain it absorbs moisture, creating a local dry area of low precipitation (a **rain shadow**, **Figure 29-5**).

Section 29.2 What Conditions Does Life Require?

- Life requires nutrients, energy, liquid water, and organism-appropriate temperatures. These resources are not distributed evenly throughout Earth's surface and limit the types of organisms that can live in different areas.

Section 29.3 How Is Life on Land Distributed?

- Temperature and liquid water are the crucial limiting factors for terrestrial ecosystems. Regions with similar climate (**biomes**) have similar vegetation as determined by temperature and water availability.
- **Tropical rain forests** are equatorial, warm, and wet. They are dominated by broadleaf evergreen trees, with animal life being primarily arboreal (**Figure 29-9**). Most nutrients are tied up in vegetation.
- **Tropical deciduous forests** are slightly less equatorial than tropical rain forests and have pronounced wet and dry seasons. Plants shed their leaves during the dry season to minimize water loss.
- **Savannahs** are extensive drought-resistant grasslands (no more than 12 inches of annual rainfall) with pronounced wet and dry seasons (**Figure 29-11**).
- **Deserts** are hot and dry, with less than 10 inches of annual rainfall (**Figure 29-13**). Desert plants are extremely drought resistant, while desert animals use behavioral and physiological mechanisms to conserve water and thermoregulate.
- A **chaparral** is a desert-like area (up to 30 inches of annual rainfall), whose climate is moderated by its proximity to the coastline (**Figure 29-16**). Chaparral vegetation consists of small trees and bushes.
- **Grasslands** are located in the center of continents and receive no more than 30 inches of annual rainfall. They are composed almost entirely of continuous grass cover (**Figures 29-17** and **29-18**).
- **Temperate deciduous forests** retain enough moisture (up to 60 inches of annual rainfall) to support trees, but seasonal freezing temperatures cause deciduous plants to drop their leaves in the fall to conserve water (**Figure 29-20**).
- **Temperate rain forests** are seasonal forests that experience moderate temperatures and an abundance of rainfall (as much as 160 inches annually). The trees of these forests are typically conifers (**Figure 29-21**).
- **Taiga** has a harsher climate than temperate deciduous forest, with long, cold winters and short growing seasons. Taiga consists almost entirely of evergreen coniferous trees (**Figure 29-22**).
- **Tundra** is a treeless region bordering the Arctic Ocean that is a frozen desert (up to 10 inches of annual precipitation). Permafrost prevents tree growth and bushes are stunted in size, but perennial flowers and lichens exist (**Figure 29-24**).

Section 29.4 How Is Life in Water Distributed?

- Energy and nutrients are the crucial limiting factors for aquatic ecosystems. Nutrients are located either in bottom sediments or along the shore, where they are washed in from the surrounding land.
- **Freshwater lakes** are diverse ecosystems that are divided into **littoral**, **limnetic**, and **profundal zones** (life zones) that correspond to specific depths (**Figure 29-25**). Lakes can be classified based on poor nutrient availability (**oligotrophic**) or rich nutrient availability (**eutrophic**, **Figure 29-26**).

- **Streams** begin at a **source** where water is provided by rain or snow. Source water is clear, oxygenated, and nutrient poor. Streams join at lower elevations, forming rivers that carry nutrients from water at higher elevations (**Figure 29-27**). Rivers enter **floodplains** on their way to lakes or oceans.
- Most oceanic life is found in shallow water (where sunlight can penetrate) near areas of **upwelling** (where nutrients are plentiful). **Coastal waters** contain the most abundant life and consist of **intertidal** and **nearshore zones**, each with their own particular autotrophic and heterotrophic organisms (**Figure 29-28**). **Coral reefs** occur in warm tropical waters and consist of specialized algae and corals that build reefs out of calcium carbonate skeletons. Coral reefs are the most diverse undersea ecosystems (**Figure 29-30**).
- Most **open ocean** life is found in the **photic zone**, where light supports phytoplankton. The **aphotic zone** is supported by nutrients that drift down from the photic zone (**Figure 29-28**).
- **Deep ocean** life is supported by excrement and dead bodies that drift down from above (**Figure 29-32**). Deep ocean life is often bioluminescent—the light is used to see, attract mates, and attract prey.
- **Hydrothermal vent communities** are supported by chemosynthetic bacteria that act as autotrophs that use hydrogen sulfide (discharged from cracks in Earth's crust) to make energy (**Figure 29-33**). Many hydrothermal vent organisms use these chemosynthetic bacteria, either as food or as symbionts to manufacture energy.

LEARNING OBJECTIVES

Section 29.1 What Factors Influence Earth's Climate?

- Explain the difference between weather and climate.
- Discuss how air currents and ocean currents are created.
- Explain the role of Earth's tilt in the seasons.
- Discuss the role of ocean currents on climate.
- Discuss the cause and effects of El Niño.

Section 29.2 What Conditions Does Life Require?

- Discuss the four requirements for life.

Section 29.3 How Is Life on Land Distributed?

- Discuss the effect water and temperature have on species distribution.
- Describe the dominant plant species and climate for each terrestrial biome.
- Discuss the impact of clear-cutting the rainforest in relationship to nutrient storage and availability.
- Discuss the role of humans in desertification.
- Discuss the impact of humans on global biodiversity.

Section 29.4 How Is Life in Water Distributed?

- Discuss the differences between water and terrestrial ecosystems.
- Describe the organisms found in the different lake-life zones.
- Discuss the impact of eutrophication on lake communities.
- Discuss the benefits provided by healthy wetlands.
- Describe the cycling of nutrients between the photic and aphotic zones of the oceans.
- Describe the habitats and organisms of the intertidal and near-shore zones of the oceans.
- Discuss humanity's impact on the intertidal and near-shore zones.
- Discuss how dead zones develop and their impact on the ocean biome.
- Describe the food web at a hydrothermal vent.

QUIZ 1

1. The Great Basin Desert in Nevada and Utah is a result of _____.
   a. a lack of drainage
   b. overgrazing
   c. a permanent high barometric pressure zone
   d. poor soil
   e. a rain shadow

2. Based on what you know about winds and air pressure zones, at which latitude would you expect to find deserts?
   a. the equator
   b. 30 degrees
   c. 40 degrees
   d. 60 degrees
   e. 80 degrees

3. Which of the following factors naturally influences Earth's climate?
   a. distribution of land and water
   b. elevation of land
   c. angle at which sunlight strikes the planet
   d. air currents
   e. all of the above

4. Every few years, the tradewinds die down, allowing warm surface waters to flow back eastward in the southern Pacific Ocean. This causes winter rains in Peru, droughts in Indonesia and South Africa, and a reduction in the anchovy harvest off the coast of Peru. This phenomenon is called _____.
   a. El Niño
   b. La Niña
   c. the Gulf Stream
   d. eutrophication
   e. upwelling

5. At about 30° north and south of the equator, there are very dry regions on Earth. Why does this occur?
   a. Cool air falls, is warmed, and absorbs moisture.
   b. Warm air falls and absorbs moisture.
   c. Cool air rises and water condenses.
   d. Warm air rises and water is evaporated.

6. The greatest diversity of plants and animals in terrestrial (land-based) environments, in terms of numbers of species is found in
   a. broad-leafed (deciduous) forests.
   b. coniferous (evergreen) forests.
   c. savannas.
   d. tropical rain forests.
   e. deserts.

7. In which terrestrial biome would you expect the most precipitation?
   a. chaparral
   b. grassland
   c. northern coniferous forest
   d. temperate deciduous forest
   e. tropical rain forest

8. The largest expanses of undisturbed and uncut forests exist in the _____.
   a. taiga
   b. tropical rain forest
   c. temperate deciduous forest
   d. temperate rain forest
   e. tropical deciduous forest

9. _____ biomes are dominated by grasses but have scattered trees and thorny bushes. They have distinct wet and dry seasons. In Africa, this biome type supports huge herds of migrating large mammals.
   a. Savanna
   b. Tropical rain forest
   c. Taiga
   d. Prairie
   e. Tundra

10. In which biome is the smallest fraction of carbon and nutrients present in the soil?
    a. tropical rain forest
    b. savanna
    c. tundra
    d. grassland
    e. coniferous forest

11. The world's deserts are located
    a. inland from major oceans or seas.
    b. 30° North or South latitude.
    c. on the leeward side of large mountain ranges such as the Sierra Nevadas.
    d. all of the above.
    e. (a) and (c).

12. Which type of climate is typical of a rain forest biome?
    a. warm temperatures with seasonal rainfall
    b. hot temperatures with little rainfall
    c. cool temperatures with heavy rainfall
    d. warm temperatures with heavy or year-round rainfall
    e. warm temperatures with high humidity and much fog

13. Extensive, shallow root systems that can quickly soak up water after infrequent rainstorms and spiny leaves and waxy coatings to reduce water loss are adaptations seen in most plants in a
    a. desert.
    b. prairie.
    c. tropical rain forest.
    d. tundra.
    e. temperate deciduous forest.

144

14. Damage from sediments, excess nutrients from logging, farming, and development on land; and rising water temperatures resulting from global warming are critically endangering which marine ecosystem?
    a. coral reefs
    b. intertidal zones
    c. estuaries
    d. open oceans
    e. hydrothermal vent communities

15. Where would you most likely have to go to encounter a geothermal vent that spews hydrogen sulfide that is ingested by sulfur bacteria?
    a. photic zone
    b. aphotic zone

QUIZ 2

1. Rising air _____; this causes water vapor to _____.
    a. heats, evaporate
    b. heats, condense
    c. cools, evaporate
    d. cools, condense

2. Mountain ranges create deserts by
    a. lifting land up into colder, drier air.
    b. completely blocking the flow of air into desert areas, thus preventing clouds from moving in.
    c. forcing air to first rise and then fall, thus causing rain on one side of the mountains and desert on the other.
    d. causing the global wind patterns that make certain latitudes very dry.
    e. causing very steep slopes that are subject to erosion.

3. The primary driver of both weather and climate is
    a. ocean currents.
    b. the sun.
    c. the ozone layer.
    d. elevation.

4. Air that is warm
    a. sinks.
    b. moves more quickly.
    c. rises.
    d. loses moisture.

5. Which of the following factors is MOST significant for explaining why Earth experiences seasons?
    a. elevation
    b. tilt of Earth
    c. distance of Earth from the sun
    d. none of the above

6. What is the primary reason that plants from distant, but climatically similar, places commonly look the same?
    a. common ancestry
    b. adaptation to the same physical conditions
    c. adaptation to similar herbivores
    d. continental drift
    e. effects of past climate change

7. Which of the following is NOT one of the four fundamental resources required for life?
    a. nutrients
    b. energy
    c. water vapor
    d. suitable temperature

8. Which of the biomes is home to the greatest concentration of biodiversity on the planet?
    a. savannas
    b. tropical deciduous forests
    c. tropical rain forests
    d. temperate deciduous forests

9. Which of these terrestrial biomes is the driest?
    a. chaparral
    b. coniferous forest
    c. deciduous forest
    d. desert
    e. grassland
    f. rain forest

10. Because of reduced evaporation resulting from fog, many coastal regions bordering deserts are characterized by small woody plants with adaptations to conserve water. What is the name of this biome?
    a. chaparral
    b. savanna
    c. steppe
    d. taiga
    e. tundra

11. In what terrestrial biome would you expect the climate to allow the most biological productivity?
    a. chaparral
    b. grassland
    c. northern coniferous forest
    d. temperate deciduous forest
    e. temperate rain forest
    f. tropical rain forest

12. Why are so many desert animals active only at night (nocturnal)?
    a. Reflection of light from the bright sun makes it too difficult for visual predators to hunt.
    b. The surface temperature of the ground would be too hot to walk on.
    c. Temperatures are lower and humidity is higher at night.

13. Which biome contains soils that are extremely rich in nutrients resulting from the accumulation of dead organic matter over many centuries?
    a. prairie
    b. tropical rain forest
    c. desert
    d. taiga
    e. temperate deciduous forest

14. Atlantic cod were once phenomenally abundant in the shallow waters off the coast of New England and the Maritime Provinces of eastern Canada. In fact, it has been said that this fishery alone fed the early European colonists to North America for the first century or more they were here. In the last half-century, however, Atlantic cod fishery has almost completely collapsed. This decline has resulted from _____.
    a. overfishing taking far more fish than could be replaced by normal reproduction
    b. acid rain causing the marine food web to collapse
    c. the death of eggs and fry as the ozone hole has increased UV damage
    d. global warming
    e. overpopulation exceeding the carrying capacity of the system for the cod, causing massive starvation

15. Which type of freshwater lake contains few nutrients?
    a. oligotrophic lakes
    b. eutrophic lakes

ANSWER KEY

Quiz 1

1. e
2. b
3. e
4. a
5. b
6. d
7. e
8. a
9. a
10. a
11. d
12. d
13. a
14. a
15. b

Quiz 2

1. d
2. c
3. b
4. c
5. b
6. b
7. c
8. c
9. d
10. a
11. f
12. c
13. a
14. a
15. a

CHAPTER 30: CONSERVING LIFE ON EARTH

OUTLINE

Section 30.1 What Is Biodiversity, and Why Should We Care About It?

- **Biodiversity** describes the diversity of organisms, their genes, and the ecosystems in which they live. **Conservation biologists** seek to preserve biodiversity.
- Ecosystem biodiversity provides many beneficial services that help sustain human life (**ecosystem services, Figure 30-1**), either directly or indirectly. Direct ecosystem services include food, building materials, medicines, natural fibers and fabrics, oxygen replenishment, and fuel. Indirect ecosystem services include soil formation, erosion control, climate regulation, genetic resources, and recreation.
- The field of **ecological economics** attempts to measure the value of ecosystem services and assess the consequences when ecosystems are damaged from human profit-making activities.

Section 30.2 Is Earth's Biodiversity Diminishing?

- **Background extinction rate** describes the natural rate of extinction in the absence of cataclysmic events. This occurs naturally at a low rate.
- **Mass extinctions** describe cataclysmic events in which many forms of life go extinct. Five mass extinctions have occurred previously and many biologists believe that human activities are now causing a sixth mass extinction.
- The World Conservation Union (IUCN) has established a list that classifies at-risk species (**Figure 30-6**).

Section 30.3 What Are the Major Threats to Biodiversity?

- Biodiversity decline is due to the increasing amount of Earth's resources used to support humanity (exceeding Earth's **biocapacity, Figure 30-7**) and the direct impact of human activity.
- Human activities that directly impact Earth include **habitat destruction** and **fragmentation** (**Figures 30-8** and **30-9**), species **overexploitation** (**Figure 30-10**), displacement of native wildlife by **invasive species** (**Figure 30-11**), pollution, and global warming.

Section 30.4 How Can Conservation Biology Help to Preserve Biodiversity?

- **Conservation biology** is an integrated science that seeks to understand the impact of human activities on natural ecosystems, to preserve and restore natural communities, to reverse the escalating loss of Earth's biodiversity, and to foster the sustainable use of Earth's resources.
- Conservation efforts seek to establish **core reserves** that are protected from human activity and connect them with **wildlife corridors**, thus promoting functional, self-sustainable communities (**Figure 30-12**).

Section 30.5 Why Is Sustainability the Key to Conservation?

- **Sustainable development** promotes long-term ecological and human well-being without compromising the future. This can be accomplished by maintaining biodiversity, recycling raw materials, and using renewable resources.
- **Biosphere reserves** provide models for conservation and sustainable development by maintaining biodiversity while preserving local cultural values (**Figure 30-14**).
- **Sustainable agriculture** helps preserve natural communities by using agricultural techniques that minimize the impact on the environment (**Table 30-1**).
- Human population growth is unsustainable and is causing resource consumption beyond Earth's biocapacity. Ultimately, human population growth must be curtailed if Earth's resources are to support life in the long term.

LEARNING OBJECTIVES

Section 30.1 What Is Biodiversity, and Why Should We Care About It?

- Describe the tangible benefits the natural environment provides.
- Discuss the resources provided directly and indirectly by healthy ecosystems.
- Discuss the economics of preserving the environment.

Section 30.2 Is Earth's Biodiversity Diminishing?

- Discuss the difference between the background extinction rate and the current rate of extinctions.

Section 30.3 What Are the Major Threats to Biodiversity?

- Discuss the difference between humanity's ecological footprint and Earth's carrying capacity.
- Discuss the combined impact of habitat destruction and fragmentation.
- Discuss the issues underlying overexploitation of resources.
- Explain the problems caused by introduction of nonnative species.
- Discuss the impact of global warming on extinction rates.

Section 30.4 How Can Conservation Biology Help to Preserve Biodiversity?

- Explain how corridors increase useable wildlife habitat.

Section 30.5 Why Is Sustainability the Key to Conservation?

- Explain the four sustainability principles.
- Describe the biosphere reserve program.
- Discuss ways people in developed countries can support sustainable practices worldwide.

QUIZ 1

1. Which of the following levels of biodiversity is directly decreased by draining a wetland to create farmland?
   a. species diversity
   b. genetic diversity
   c. ecosystem diversity
   d. none of the above

2. Plants have been a huge source of medicines for humans. For example, extracts of rosy periwinkle have provided medicine for treating cancer. The ability to discover future medicines like these from plants is a strong argument for
   a. genetic engineering.
   b. overexploitation.
   c. preserving biodiversity.
   d. supporting pharmaceutical industries.

3. Which of the following is NOT a benefit that humans receive from a biologically diverse world?
   a. abundant water
   b. economic gains
   c. medicine
   d. food

4. Which of the following is NOT true about extinction of species?
   a. Extinction is a natural process.
   b. Extinctions occur at a relatively low rate.
   c. Mass extinctions have occurred in the past.
   d. The current rate of extinction is equal to background rates.

5. We are currently experiencing what scientists consider a biodiversity crisis. What is the main cause of the currently high extinction rate?
   a. human activity
   b. climate change
   c. volcanic activity
   d. increased solar radiation resulting from a diminished ozone layer

6. The current rate of extinction is estimated to be _____ times higher than background extinction rates.
   a. 1–10
   b. 10–100
   c. 100–1000
   d. 1000–10,000

7. Your personal "ecological footprint" depends on
_____.
   a. the size of your house, how well it is insulated, and the climate in the region where you live
   b. the types of transportation you use, how much you travel, and whether you travel alone or with a group
   c. the types and amounts of foods you eat and how much you waste
   d. how many goods you consume and the waste you produce
   e. all of the above

8. Which process is a major threat to biodiversity?
   a. climate change
   b. invasive species
   c. habitat alteration
   d. all of the above

9. A recent study estimates that at the present rate of harvest, the world's fisheries will be depleted by 2050. This is an example of
   a. pollution.
   b. overexploitation.
   c. habitat destruction.
   d. competition.

10. Which of the following is NOT a goal of conservation biologists?
   a. understand the negative impact of animal habitats on human activity
   b. preserve and restore natural communities
   c. reverse the escalating loss of Earth's biodiversity caused by human activities
   d. foster sustainable use of Earth's resources

11. Conservation biology depends on the support and expertise of
   a. people in a wide array of fields in biology.
   b. government officials.
   c. educators.
   d. individuals.
   e. all of the above.

12. In Yellowstone Park, the removal of wolves had a wide-reaching negative effect on the level of biodiversity. Wolves are predators of elk and keep their populations in check. Without the wolves, elk eat the young aspens, which means that aspen groves could not regenerate. Aspen groves provide shelter and homes for many species of plant and bird, as well as material for beavers to build dams. The dams create marshlands that in turn create habitat for many aquatic organisms. Since the biodiversity of these communities depends on the presence of wolves, they are considered to be
   a. dominant species.
   b. keystone species.
   c. domino species.
   d. predators.

13. Which of the following is NOT a feature of sustainable development?
   a. diverse communities with a richness of community interactions
   b. relatively stable populations that remain within the carrying capacity of the environment
   c. recycling and efficient use of raw materials
   d. reliance on nonrenewable sources of energy

14. Biosphere Reserves around the world
   a. adhere to the ideal Biodiversity Reserve model set down by the United Nations.
   b. are voluntary locally managed by the country where they are located.
   c. are adequately funded.
   d. all of the above

15. No-till farming is an example of sustainable agriculture because it.
   a. uses natural herbicides and pesticides to kill plant and animal pests.
   b. leaves remains of harvested plants behind to form mulch for the next year's plants.
   c. reduces soil erosion.
   d. all of the above

QUIZ 2

1. Which of the following is an argument for protecting biodiversity?
   a. Decreasing biodiversity can result in a loss of resources that could provide humans with new drugs to cure disease or that contain genes that would produce better strains of crops.
   b. Organisms are linked in a complex web of life. Disrupting natural ecosystems may have unintended consequences; for example, insecticides have killed many of the pollinators of our agricultural crops.
   c. A hypothesis put forth by E.O. Wilson, biophilia, suggests that humans enjoy and derive benefit from natural landscapes.
   d. all of the above

2. Which of the following is an ecosystem service provided free of charge by Earth's diverse ecosystems?
   a. soil production
   b. plant pollination
   c. water production
   d. erosion control
   e. all of the above

3. Species with several alleles for many of their genes display _____ genetic diversity and are _____ likely to be able to adapt in response to changing conditions.
   a. high, more
   b. high, less
   c. low, more
   d. low, less

4. Many of the species that are thought to be going extinct each year because of human activity do not appear on the lists of extinct species. Why?
   a. Many species are unknown to science and probably go extinct before they have been described.
   b. There really are not that many species that have become extinct.
   c. Many species are not listed because they are not considered important.
   d. Only extinctions that are naturally caused are listed.

5. If the present rate of extinction continues this will be the _____ mass extinction in history.
   a. second
   b. third
   c. fourth
   d. fifth
   e. sixth

6. Which of the following would be the hardest groups in which to document extinction rates?
   a. mammals
   b. birds
   c. insects
   d. all are equally hard

7. Rerouting the Mississippi River for commercial reasons has played a role in the loss of wetlands in southern Louisiana. In addition to the loss of territory for many species, this also removed an important buffer to storm surges, which led to the disastrous flooding of New Orleans from hurricane Katrina in 2004. This is an example of the consequences of
   a. pollution.
   b. overexploitation.
   c. habitat destruction.
   d. competition.

8. The zebra mussel was accidentally introduced into U.S. waterways from the ballast water released from ships. It reproduces much more successfully than native mussels and is out-competing native aquatic organisms. The zebra mussel is an example of a(an)
   a. introduced species.
   b. parasite.
   c. predator.
   d. keystone species.

9. Global climate change is likely to increase the rate of extinction worldwide. The pollutant most implicated in this problem is
   a. nitrous oxides.
   b. carbon dioxide.
   c. sulfur dioxides.
   d. chloroflurocarbon.

10. Which of the following is a statement that you would expect to hear from a conservation biologist?
    a. Protecting the natural environment is crucial for protecting human welfare.
    b. The elaborate relationships among organisms that have evolved over millennia should be preserved within their natural environments.
    c. Other species should be protected even if they have no value to humans.
    d. All of the above are correct.

11. Core reserves
    a. are vital stockpiles of fossil fuels being reserved for future energy use.
    b. are vital reservoirs of freshwater for future use.
    c. are areas set aside for human use and recreation.
    d. are areas that encompass entire ecosytems and their biodiversity that are protected from all but very low-impact human activity.

12. Based on the burgeoning human population, today's individual core reserves are seldom large enough to meet the minimum required areas for many species. If you were in charge of core reserves in a region of the country, which of the following would you choose as the most efficient strategy for solving this problem?
    a. spearheading human-population reduction campaigns urging couples to have only one or two children
    b. building wildlife corridors to connect reserves, thus enlarging the overall habitat range
    c. buying and developing land around the existing park to enlarge the area
    d. none of the above

13. Which of the following is NOT an example of sustainable development?
    a. the present commercial harvest of fish
    b. harvesting trees from forests where logging and tree replanting and regrowth are balanced
    c. using wind for energy
    d. using biofuels for energy

14. A Biosphere Reserve is all of the following EXCEPT a(an)
    a. core reserve that is protected from development and sees only low-impact human activities.
    b. inner zone of the core reserve that is protected from all human activity.
    c. surrounding buffer zone that permits low-impact human activities and development.
    d. transition area that supports settlements, tourism, fishing, and agriculture, operated sustainably.

15. Why is shade-grown coffee an example of a sustainable crop?

    a. Since it grows in the shade, it can be grown in tracts of relatively undisturbed rain forest.

    b. The trees of the remaining forest create a habitat for a large diversity of wildlife.

    c. The trees of the remaining reduce soil erosion.

    d. all of the above

ANSWER KEY

Quiz 1

1. c		**9.** b
2. c		**10.** a
3. a		**11.** e
4. d		**12.** b
5. a		**13.** d
6. c		**14.** b
7. e		**15.** d
8. d		

Quiz 2

1. d		**9.** b
2. c		**10.** d
3. a		**11.** d
4. a		**12.** b
5. e		**13.** a
6. c		**14.** b
7. c		**15.** d
8. a		

CHAPTER 31: HOMEOSTASIS AND THE ORGANIZATION OF THE ANIMAL BODY

OUTLINE

Section 31.1 Homeostasis: How Do Animals Regulate Their Internal Environment?

- **Homeostasis** describes the dynamic equilibrium within the animal body by which physiological conditions are maintained.
- Animals regulate body temperature by deriving body warmth from the environment (**ectotherms**) or from metabolic activities within their own bodies (**endotherms**). Ectotherms can usually tolerate greater body temperature extremes than **endotherms** (**Figure 31-1**).
- Homeostasis is most often maintained by **negative feedback** (**Figure 31-2**), in which a body change causes a response that counteracts it and returns conditions back to normal (the **set point**).
- Sometimes **positive feedback** occurs, in which a change initiates events that intensify that change. An example is the control of uterine contractions during childbirth.

Section 31.2 How Is the Animal Body Organized?

- **Cells** are the building blocks of all life.
- **Tissues** are made of cells of similar structure that perform a specific function. Animal tissues include **epithelial**, **connective**, **muscle**, and **nerve tissue**. **Epithelial tissue** forms membranous coverings over internal and external body surfaces and also forms **glands** (**Figure 31-4**). **Connective tissue** is composed of cells and extracellular matrix that usually serve to bind other tissues. **Loose connective tissue** includes the dermis, fibrous **connective tissue** includes tendons and ligaments, and **specialized connective tissue** includes cartilage, bone, adipose, blood, and lymph (**Figures 31-5** to **31-8**). **Muscle tissue** has the ability to contract. **Skeletal muscle** is under voluntary control and produces skeletal movements (**Figure 31-9**). **Smooth muscle** is under involuntary control and commonly found in tubular organs. **Cardiac muscle** is under involuntary control and causes heart contractions. Nerve tissue allows you to sense and respond to stimuli by the transmission of electrical impulses. Nerve tissue is composed of cells that transmit electrical impulses (**neurons**) and those that support them (**glial cells**).
- **Organs** are made of more than one tissue type and perform complex functions. An example is the **skin** (**Figure 31-11**).
- **Organ systems** are groups of two or more organs that work together for a common function. Organ systems include the circulatory, lymphatic/immune, digestive, urinary, respiratory, endocrine, nervous, muscular, skeletal, and reproductive systems (**Table 31-1**).

LEARNING OBJECTIVES

Section 31.1 Homeostasis: How Do Animals Regulate Their Internal Environment?

- Describe the functions regulated by homeostasis.
- Explain the costs and benefits of endothermic versus ectothermic temperature regulation.
- Explain the difference between negative feedback and positive feedback and their uses in the body.

Section 31.2 How Is the Animal Body Organized?

- Describe the organization of animals from cell to organism.
- Describe the structure and functions of epithelial tissues.

- Describe the structure and functions of endocrine glands and exocrine glands.
- Describe the structure and functions of connective tissues.
- Describe the structure and functions of the three types of muscle tissue.
- Describe the structure and functions of nervous tissues.
- Describe the interaction of the different tissue types that make up the skin.
- Describe the basic structures and functions of the organ systems in the human body.

QUIZ 1

1. The constancy of an animal body's internal environment is maintained by _____.
   a. a single feedback mechanism
   b. a few independent feedback systems
   c. a coordinated, integrated network of systems

2. Special sensory nerve endings in the skin of the hand are responsive to temperature. When an extremely hot object is encountered, nerves conduct this information to the spinal cord, which, in turn, sends a signal to skeletal muscle, causing it to contract and pull the affected part of the body away from the stimulus (often before the sensation of a burn is felt).

   In this scenario, the control center is _____.
   a. the nerve endings in the skin
   b. the spinal cord
   c. the skeletal muscles of the hand
   d. the nerves conducting impulses from the sensory nerves to the spinal cord

3. Fish do not maintain whole-body temperatures different from the temperature of the water in which they live. However, many fish, if given a choice of water temperatures (say, in an experimental aquarium that offers a gradient of water temperatures), will select a narrow range of water temperatures in which to live. Thus, they exhibit a "preferred temperature" that they can maintain by controlling the amount of time they spend in water of different temperatures.

   Does this represent true homeostasis in the fullest sense of the term? Why?
   a. Yes, because a constant body temperature is maintained.
   b. Yes, because body temperature is actively regulated such that internal physiological variables are kept within the range that cells need to function.
   c. No, because, even though the fish are maintaining relatively constant internal conditions, they are not using a feedback system in order to maintain these conditions.
   d. No, because homeostasis involves the control of a physiological variable within very narrow limits so that cells can function. The body temperature of a fish fluctuates with the temperature of the external environment.

4. How do animals regulate their physiology so that the physiological parameters (e.g., regulation of pH, body temperature, electrolyte balance) stay within narrow limits?
   a. homeostasis
   b. dynamic equilibrium
   c. negative feedback systems
   d. all of the above
   e. none of the above

5. Consider a scenario in which your core body temperature drops to 91°F. If a negative feedback mechanism responded to this change, what would it do to that core temperature?
   a. It would drop the body temperature further toward a negative, or colder, value.
   b. There would be no change in the body temperature due to negative feedback.
   c. It would reverse the drop in temperature so that the body begins to warm toward its normal operating temperature.
   d. None of the above is correct.

6. Pretend that your body would respond to a decrease in body temperature using positive feedback. In this hypothetical example, if body temperature were to drop from 98.6°F to 97°F, positive feedback would work to
   a. continue to lower body temperature.
   b. reverse the dropping temperature and begin to warm the body again.
   c. maintain temperature at 97°F with no further changes.
   d. None of the above

7. In an unfortunate accident, a young man is struck in the leg by a bullet that tears through the femoral artery just above his right knee. The subsequent blood loss leads to a rapid drop in blood pressure. Baroreceptors in the aorta monitor blood pressure and send signals to the vasomotor center of the brain. In this case, the vasomotor center will increase its sympathetic stimulation of the blood vessels, which will cause them to constrict. This constriction leads to an increase in blood pressure. This is a description of _____.
   a. a positive feedback system
   b. a negative feedback system
   c. both positive and negative feedback systems

153

8. In the situation described in question 7, what is the variable, or physiological parameter, that this system controls?
   a. blood pressure
   b. blood volume
   c. blood vessel diameter
   d. sympathetic nerve activity

9. In the situation described in question 7, which of the following represents the control center in this feedback system?
   a. blood pressure
   b. vasomotor center
   c. blood vessel diameter
   d. baroreceptors

10. In the situation described in question 7, which of the following represents the effector in this feedback system?
    a. blood vessel diameter
    b. baroreceptors
    c. vasomotor center
    d. blood pressure

11. The skin contains
    a. epithelial tissue.
    b. connective tissue.
    c. nerve tissue.
    d. muscle tissue.
    e. all of the above.

12. Glands that become separated from the epithelium that produced them are called _____ glands.
    a. sebaceous
    b. sweat

c. exocrine
d. endocrine
e. saliva

13. All of the following are examples of connective tissue EXCEPT
    a. tendons
    b. ligaments
    c. blood
    d. muscle
    e. adipose tissue

14. Which of the following statements about muscle is true?
    a. Smooth muscle is important in locomotion.
    b. Skeletal muscle is not under conscious control.
    c. Cardiac muscle utilizes gap junctions.
    d. Smooth muscle is called voluntary muscle.
    e. Smooth muscle moves the skeleton.

15. All of the following are found in the dermis EXCEPT
    a. arteries.
    b. sensory nerve endings.
    c. hair follicles.
    d. sebaceous glands.
    e. cells packed with keratin.

QUIZ 2

1. Homeostasis is the condition in which the body maintains
   a. a low level of energy usage.
   b. a relatively constant internal environment.
   c. a body temperature without limits.
   d. a stable external environment in which to live.

2. Which of the following statements about feedback systems is TRUE?
   a. Blood sugar levels are regulated by positive feedback systems.
   b. Negative feedback systems work to increase the original stimulus in the feedback system.
   c. Positive feedback systems are always harmful because the original stimulus runs "out of control."
   d. Negative feedback systems work to keep physiological variables within limits around a set point.

3. Dynamic equilibrium refers to the body's ability to
   a. maintain a stable internal environment that is allowed to fluctuate within limits.
   b. maintain an internal environment that does not change.
   c. work without limits.
   d. alter the external environment to make it a more suitable place to live.

4. In a negative feedback system, an input stimulus triggers the body to create an output response that
   a. enhances the changes caused by the input stimulus.
   b. decreases the set point.
   c. reverses the changes caused by the input stimulus.
   d. blocks the input stimulus.

5. In a positive feedback system, an input stimulus triggers the body to create an output response that
   a. enhances the changes caused by the input stimulus.
   b. decreases the set point.
   c. reverses the changes caused by the input stimulus.
   d. blocks the input stimulus.

6. When a snowball rolls down a snowy hill, it picks up snow, which causes it to roll faster. The result is that the snowball will pick up more snow and roll even faster. This scenario would best be described as
   a. a positive feedback system.
   b. a negative feedback loop.
   c. dynamic equilibrium.

7. The enhancement of uterine labor contractions by oxytocin would be considered a
   a. negative feedback system.
   b. dynamic equilibrium.
   c. positive feedback system.

8. Which of the following represents the role of the effector in a feedback system?
   a. It contains the set point for the variable that is being controlled.
   b. It measures changes in the variable that is being controlled.
   c. It produces output that will then change the variable that is being controlled.

9. Which of the following represents the role of the control center in a feedback system?
   a. It contains the set point for the variable that is being controlled.
   b. It measures changes in the variable that is being controlled.
   c. It produces output that will have a direct effect on the variable that is being controlled.

10. Desert lizards rely on energy from the sun to regulate their body temperatures. Heat from the sun penetrates the skin and warms the blood, which is then circulated to the body core and other regions, warming them. When desert lizards need to cool their bodies, they move out of the sunlight until body temperatures drop. They must also, of course, be able to maintain water balance in extremely dry environments. Which of the following features would you NOT expect to find in reptilian skin?
    a. sweat glands
    b. a thick layer of heavily keratinized cells
    c. a relatively thin dermis

11. Which of the following is NOT one of the major categories of animal tissue?
    a. endocrine tissue
    b. connective tissue
    c. epithelial tissue
    d. muscle tissue
    e. nerve tissue

12. Which of the following is NOT a type of connective tissue?
    a. fat
    b. blood
    c. bone
    d. tendons
    e. glands

13. Organs are _____.
    a. formed of all four tissue types
    b. formed of two or more tissues that operate independently
    c. formed of two or more tissues that function together

14. Which of the following represents the organizational levels that make up the body of an animal from the most simple to the most complex?
    a. cell—tissue—organ—organ system
    b. organ system—organ—tissue—cell
    c. tissue—cell—organ system—organ
    d. tissue—organ—cell—organ system

15. Which type of connective tissue are tendons and ligaments?
    a. loose connective tissue
    b. fibrous connective tissue
    c. specialized connective tissues
    d. none of the above

ANSWER KEY

Quiz 1

1. c
2. b
3. b
4. d
5. c
6. a
7. b
8. a

9. b
10. a
11. e
12. d
13. d
14. c
15. e

Quiz 2

1. b
2. d
3. a
4. c
5. a
6. a
7. c
8. c

9. a
10. a
11. a
12. e
13. c
14. a
15. b

CHAPTER 32: CIRCULATION

OUTLINE

Section 32.1 What Are the Major Features and Functions of Circulatory Systems?

- All circulatory systems are made of **blood**, **blood vessels**, and at least one **heart**.
- **Open circulatory systems** consist of at least one heart that pumps blood through vessels into a **hemocoel**, where blood bathes internal tissues and organs (**Figure 32-1a**). Most invertebrates have this type of circulatory system.
- **Closed circulatory systems** consist of at least one heart and blood vessels (**Figure 32-1b**). Blood never leaves the blood vessels and is efficiently transported to tissues and organs because of the higher blood pressures. All vertebrates and some invertebrates have this type of circulatory system.
- The vertebrate circulatory system transport nutrients, hormones, gases, and wastes throughout the body. It also regulates temperature, forms clots to prevent blood loss, and protects the body against bacteria and viruses.

Section 32.2 How Does the Vertebrate Heart Work?

- The vertebrate heart consists of **atria**, which collect blood from the body, and **ventricles**, which send blood to the body tissues.
- The first vertebrate heart that evolved is seen in fish and is two chambered (one atrium and one ventricle). As fish gave rise to amphibians and reptiles, a three-chambered heart evolved (two atria and one ventricle) that allowed for partial separation between oxygenated and deoxygenated blood. As reptiles gave rise to mammals and birds, a four-chambered heart evolved (two atria and two ventricles) that allowed for the complete separation of oxygenated blood and deoxygenated blood (**Figure 32-2**).
- **Veins** carry blood to the heart, depositing it in the atria of the heart. When ventricles contract, blood is carried away from the heart by **arteries**. Deoxygenated blood enters the right atrium, which contracts and forces blood into the right ventricle. The right ventricle contracts and forces blood through pulmonary arteries to the lungs, where it is oxygenated. Oxygenated blood enters the left atrium through pulmonary veins. This atrium contracts and forces blood into the left ventricle. The left ventricle contracts and forces blood through the aorta (a major artery) to the rest of the body (**Figure 32-3**).
- **Cardiac muscle** is composed of cells joined by intercalated discs that allow electrical signals to spread among cells and cause interconnected heart regions to contract almost synchronously (**Figure 32-4**).
- The cardiac cycle consists of atrial contraction, followed by ventricular contraction and a brief period of atrial and ventricular relaxation (**Figure 32-5**). **Systolic pressure** is generated during ventricular contraction, while **diastolic pressure** occurs during ventricular relaxation (**Figure 32-6**).
- One-way heart valves maintain the direction of blood flow through the heart (**Figures 32-3** and **32-5**).
- Heart contractions are initiated by electrical impulses generated by the **sinoatrial node**. The contractions are then coordinated by the **atrioventricular node** and **Purkinje fibers** (**Figure 32-7**).
- The nervous system and hormones modify heart rate.

Section 32.3 What Is Blood?

- **Blood** is composed of fluid **plasma** and cellular components (**red blood cells**, **white blood cells**, and **platelets**).
- **Plasma** is composed of water, proteins, salts, nutrients, and wastes.
- Red blood cells (**erythrocytes**) contain iron-containing proteins (**hemoglobin**) that carry oxygen (**Figure 32-9**). Red blood cell production occurs in the bone marrow and is influenced by the hormone erythropoietin (**Figure 32-10**).
- White blood cells (**leukocytes**) come in five types and are specialized to fight infection and disease (**Figure 32-11**).
- Platelets are fragments of cells (**megakaryocytes**) that stick to damaged blood vessels and release substances that result in the formation of sticky protein threads (**fibrin**) that form a blood clot (**Figure 32-12**).

Section 32.4 What Are the Types and Functions of Blood Vessels?

- Once blood leaves the heart, it travels sequentially through arteries, arterioles, capillaries, venules, and veins (**Figure 32-14**). Veins return blood to the heart.

- **Arteries** are vessels with thick muscular walls with elastic properties that recoil during ventricular relaxation, thus maintaining blood flow.
- **Arterioles** are small-diameter muscular vessels that control the distribution of blood flow by constricting, or by controlling **precapillary sphincter muscles**. Arteriole diameter is controlled by the sympathetic nervous system and local factors, while precapillary sphincters are controlled by local factors.
- **Capillaries** receive blood from arterioles. They are microscopic and have thin walls that allow for the exchange of nutrients and wastes between cells and the blood (**Figure 32-15**).
- **Venules** collect capillary blood and deposit it into larger **veins**. Veins are wide and provide a low-resistance pathway for blood flow. Veins contain one-way valves that maintain blood flow in spite of the low blood pressure in veins (**Figure 32-16**).

Section 32.5 How Does the Lymphatic System Work with the Circulatory System?

- The lymphatic system includes the **lymphatic vessels**, **lymph nodes**, **tonsils**, **thymus**, and **spleen**. It functions to transport small intestine fats to the bloodstream, returns excess interstitial fluid to the bloodstream, and defends the body against bacteria and viruses.
- Lymph capillaries collect excess interstitial fluid (leaked from blood capillaries) through one-way valves. The fluid then enters lymph vessels that deposit it back into the bloodstream (**Figure 32-18**). This collected fluid is called **lymph**.
- The **tonsils**, **lymph nodes**, and **thymus** house white blood cells that cleanse the lymph of bacteria and viruses that are delivered to them by the lymph vessels. The **spleen** performs a similar task, except it cleanses blood instead of lymph.

LEARNING OBJECTIVES

Section 32.1 What Are the Major Features and Functions of Circulatory Systems?

- Describe the three parts of a circulatory system.
- Explain the difference between open circulatory and closed circulatory systems.
- Describe the functions of the vertebrate circulatory system.

Section 32.2 How Does the Vertebrate Heart Work?

- Describe the differences among fish, amphibian, and mammalian hearts.
- Describe the path of blood through a mammalian body.
- Discuss the unique attributes of cardiac muscle tissue that provide for its specialized function.
- Describe the structure and function of heart valves.
- Describe the path of electrical impulse resulting in heart muscle contraction.

Section 32.3 What Is Blood?

- Describe the components of blood and their function.

Section 32.4 What Are the Types and Functions of Blood Vessels?

- Describe the structural and functional differences between veins and arteries.
- Discuss how gas exchange takes place at capillaries.
- Discuss how and why arterioles change in diameter.

Section 32.5 How Does the Lymphatic System Work with the Circulatory System?

- Describe the structures and functions of the lymphatic system.

QUIZ 1

1. The _____ is an organism with an open circulatory system.
   a. snail
   b. sponge
   c. octopus
   d. frog
   e. human

2. Which of the following is NOT a part of all circulatory systems?
   a. a fluid that acts to transport substances throughout the body
   b. a system of passageways that carry the fluid
   c. an open region inside the body where internal organs are immersed directly in the fluid
   d. a muscular pump for pushing the fluid through the passageways

3. The muscular chambers in a heart are called _____.
   a. capillaries
   b. atria
   c. partitions
   d. ventricles
   e. (b) and (d)

4. Which of the following represents the highest blood pressure during ventricular relaxation?
   a. systolic pressure of 130
   b. blood pressure of 110/70
   c. diastolic pressure of 90
   d. blood pressure of 100/80

5. The evolution of the cardiovascular system in vertebrates involved several important changes. Which one of the following represents the order of a major change that occurred?
   a. a closed to an open circulatory system
   b. three-chambered to a two-chambered and then a four-chambered heart
   c. four-chambered to a three-chambered and then a two-chambered heart
   d. an open to a closed circulatory system

6. Mixing of oxygenated and deoxygenated blood within the heart occurs in the _____.
   a. bird
   b. frog
   c. dog
   d. fish
   e. flatworm

7. The cardiac cycle is _____.
   a. the time period between the two heart sounds
   b. the action of the heart in a minute's time

c. the blood leaving the heart and returning to the heart
d. the synchronous contraction of the two atria followed by the contraction of the two ventricles

8. Which statement BEST describes the role of the heart in a circulatory system?
   a. It acts as a reservoir for the storage of blood.
   b. It acts as a pump to move blood through the circulatory system.
   c. It exchanges oxygen and carbon dioxide with the outside air.
   d. It is where the blood cells are produced for the circulatory system.

9. Oxygenated blood is blood that has
   a. lost some of its oxygen after dumping it off at the lungs or gills.
   b. had its oxygen levels restored after passing through the body's tissues.
   c. lost some of its oxygen to the body's tissues.
   d. had its oxygen levels restored after picking up oxygen at the lungs.

10. Which event initiates blood clotting?
    a. contact with an irregular surface by platelets and other factors in plasma
    b. production of the enzyme thrombin
    c. conversion of fibrinogen into fibrin
    d. conversion of fibrin into fibrinogen
    e. excess flow of blood through a capillary

11. Which of the following is NOT a component of plasma?
    a. water
    b. globulins
    c. fibrinogen
    d. albumins
    e. platelets

12. Erythroblastosis fetalis is a disease in newborns which is caused by _____.
    a. low blood oxygen level
    b. the mother's immune system recognizing fetal Rh factor and making Rh antibodies
    c. fetal Rh antibodies attacking maternal red blood cells with Rh factor
    d. lack of Rh proteins in the fetal blood
    e. the clotting mechanism

13. How are capillaries specialized for the exchange of respiratory gases?
    a. Capillary walls are only one cell thick.
    b. Capillary walls are wrapped in a layer of smooth muscle.

c. Capillaries allow the passage of fluids through spaces between the cells.
d. Capillaries have one-way valves.
e. (a) and (c) are correct.

14. The most important factor in the return of blood flow to the heart is _____.
a. the pumping of the heart
b. high pressure

c. skeletal muscle contraction
d. valves in the veins

15. Lymph most closely resembles
a. blood.
b. urine.
c. plasma.
d. interstitial fluid.

QUIZ 2

1. The main function of the vertebrate circulatory system is to
a. transport substances.
b. cool the body.
c. regulate the pH of fluids in the body.
d. fight against invading pathogens.

2. Which of the following is NOT an important function of the vertebrate circulatory system?
a. transport of nutrients and respiratory gases
b. regulation of body temperature
c. protection of the body by circulating antibodies
d. removal of waste products for excretion from the body
e. defense against blood loss, through clotting

3. The wall of the left ventricle is thicker than the wall of the right ventricle in order to
a. allow for a larger blood volume.
b. pump blood with greater pressure.
c. send blood to the left atrium under pressure.
d. pump blood through a smaller valve.

4. Which of the following events of the cardiac cycle occurs when systolic pressure is being measured in blood pressure?
a. The atria and ventricles are relaxed and filling passively.
b. The atria are relaxed and the ventricles are contracting.
c. All of the chambers are contracting.

5. Deoxygenated blood is delivered to the heart through the
a. pulmonary artery.
b. aorta.
c. pulmonary veins.
d. superior and inferior vena cavae.

6. Blood that flows through the pulmonary veins will be carried to the
a. left ventricle.
b. right ventricle.
c. left atrium.
d. right atrium.

7. The pulmonary semilunar valve opens when
a. the ventricles relax.
b. the atria contract.

c. both the atria and ventricles are relaxed.
d. the ventricles contract.

8. Which of the following blood pressures has the highest *systolic* reading?
a. 120/80
b. 140/70
c. 90/55
d. 150/90
e. 130/95

9. Epinephrine is given as a treatment for shock victims because it _____.
a. is an anticoagulant
b. eliminates excess water and salts from the body, reducing blood pressure
c. increases heart rate and blood pressure
d. dilates the coronary artery carrying oxygenated blood to the heart muscle

10. Why do people with kidney damage, particularly those who are on hemodialysis, tend to be severely anemic?
a. The damaged kidneys are destroying red blood cells.
b. There is a lack of erythropoietin.
c. The bone marrow cannot get enough cholesterol to make the necessary red blood cells.
d. Excessive urea excreted by the kidneys into urine prevents the bone marrow from making blood cells.

11. What is the determining factor that stimulates erythropoietin release from the kidney?
a. too much carbon dioxide in the blood
b. too little carbon dioxide in the blood
c. too much oxygen in the blood
d. too little oxygen in the blood

12. What would happen if a person received a drug that acted as a thrombin inhibitor?
a. Heart rate would drop.
b. Heart rate would increase.
c. Anemia would result because of a decrease in the number of red blood cells.
d. The ability to produce blood clots would be inhibited.

13. The liver is responsible for producing a variety of proteins found in plasma. When a person suffers from liver failure, a common symptom is edema (swelling) of the tissues. Why does this occur?
    a. The person is making smaller proteins that diffuse into the tissues. Water then follows these proteins into the tissues.
    b. The person is making fewer plasma proteins, but the few that remain are diffusing into the tissues. Water then follows these proteins into the tissues.
    c. The liver is overproducing plasma protein. The excess protein displaces water and forces it into the tissues.
    d. The liver is making fewer plasma proteins and a weak osmotic gradient is produced. Water that has entered the tissue begins to accumulate there.

14. What is the difference between arteries and veins?
    a. Arteries carry oxygenated blood; veins carry deoxygenated blood.
    b. Arteries carry blood away from the heart; veins carry blood toward the heart.
    c. Arteries carry blood toward the heart; veins carry blood away from the heart.

15. The _____ is NOT part of the lymphatic system.
    a. thyroid
    b. thymus
    c. spleen
    d. tonsils

ANSWER KEY

Quiz 1

1. a
2. c
3. e
4. c
5. d
6. b
7. d
8. b
9. d
10. a
11. e
12. b
13. e
14. c
15. d

Quiz 2

1. a
2. d
3. b
4. b
5. d
6. c
7. d
8. d
9. c
10. b
11. d
12. d
13. d
14. b
15. a

CHAPTER 33: RESPIRATION

OUTLINE

Section 33.1 Why Exchange Gases?

- Breathing and gas exchange support cellular respiration.
- When air is inhaled, O_2 is deposited into the blood and is carried to cells throughout the body for use in cellular respiration.
- Cellular respiration produces CO_2, which leaves cells and is deposited in the blood. It is then carried to the lungs where it is released from the body.

Section 33.2 What Are Some Evolutionary Adaptations for Gas Exchange?

- Gas exchange occurs by diffusion across moist, thin respiratory surfaces that have a large surface area.
- Some animals that live in moist environments lack a specialized respiratory system, instead using their small or thin and flattened bodies as respiratory surfaces (**Figure 33-1**). Animals with low metabolic demands and/or well-developed circulatory systems may also lack specialized respiratory structures.
- Larger, more active animals have developed respiratory systems that facilitate gas exchange by alternating **bulk flow** and diffusion (**Figure 33-2**).
- Many aquatic animals use **gills**, which are thin projections from the body surface into the surrounding water (**Figures 33-3** and **33-5**). Gills are often elaborately branched or folded to maximize respiratory surface area. Some animals use countercurrent exchange mechanisms to maximize the amount of O_2 extracted from the water (**Figure E33-1**).
- Terrestrial animals typically use one of two forms of internal respiratory structures: **tracheae** or **lungs**.
- Insects use **tracheae**, which are a series of internal branching tubes that transport air from the outside environment (through **spiracle** openings) directly to the cells for gas exchange (**Figure 33-4**).
- Most terrestrial vertebrates use **lungs**, which are protected internal chambers containing moist respiratory surfaces. Amphibians supplement lung breathing with the use of larval gills and cutaneous respiration (**Figure 33-5 a, b**). Reptiles, which possess water-impermeable scales, rely solely on their well-developed lungs for respiration (**Figure 33-5c**). Birds use lungs that are composed of hollow, thin walled tubes (**parabronchi**) that allow the continuous movement of air through them. This allows for the uninterrupted exchange of gases during both inhalation and exhalation, unlike in the lungs of other vertebrates (**Figure 33-6**)

Section 33.3 How Does the Human Respiratory System Work?

- The human respiratory system is composed of two parts: a **conduction portion** and a **gas-exchange portion**.
- The **conduction portion** of the human respiratory system consists of passageways that carry air in and out of the lungs (the gas-exchange portion). It is composed of the nose and mouth, pharynx, larynx, trachea, bronchi, and bronchioles (**Figure 33-7a**).
- The **gas-exchange portion** of the human respiratory system exchanges gases between the air in microscopic lung **alveoli** and the blood (**Figures 33-7b** and **33-9**). Specifically, O_2 diffuses into the blood from alveolar air and CO_2 diffuses out of the blood into alveolar air.
- Most of the O_2 in the blood is transported bound to hemoglobin within red blood cells. CO_2 is transported in the blood dissolved in the plasma, bound to hemoglobin, or as bicarbonate ions (**Figure 33-10**).
- **Breathing** involves drawing air into the lungs (**inhalation**) because of chest cavity expansion resulting from the contraction of the diaphragm and rib muscles. Relaxation of these muscles causes them to recoil, which expels air (**exhalation**) because of the partial collapse of the chest cavity (**Figure 33-11**).
- Breathing rate is controlled by the **respiratory center** of the brain, which is influenced by the concentration of respiratory gases (primarily CO_2) within the blood.

LEARNING OBJECTIVES

Section 33.1 Why Exchange Gases?

- Explain why we require oxygen and produce carbon dioxide.

Section 33.2 What Are Some Evolutionary Adaptations for Gas Exchange?

- Describe the three requirements for diffusion of CO_2 and O_2.
- Describe the paths of O_2 and CO_2.
- Explain the difference between diffusion and bulk flow of gasses.
- Discuss the structural and functional differences among gills, tracheae, and lungs.

Section 33.3 How Does the Human Respiratory System Work?

- Describe the functions of the conducting portion of the respiratory system.
- Discuss how the structure of the alveoli relate to their function.
- Explain the difference between CO_2 transport and O_2 transport.
- Explain which muscles are involved in inhalation and exhalation.
- Discuss the homeostatic mechanisms controlling breathing rate.

QUIZ 1

1. Why do animal tissues produce carbon dioxide?
   a. It is an essential nutrient.
   b. It is needed to attract oxygen.
   c. It is a waste product from aerobic cellular respiration.
   d. Carbon dioxide is not produced; it diffuses into the tissues from the blood.

2. Why is breathing so important?
   a. A fresh supply of oxygen must be constantly provided so that cells are able to metabolize glucose efficiently.
   b. The carbon dioxide constantly produced by cells must be removed from the body.
   c. Because respiration, also called *breathing,* is synonymous with cell respiration.
   d. all of the above are correct.
   e. (a) and (b)

3. After blood passes through the body's tissues, it will become
   a. enriched in oxygen and carbon dioxide.
   b. enriched with carbon dioxide and its oxygen content will decrease.
   c. enriched in oxygen and depleted of carbon dioxide.
   d. depleted of both oxygen and carbon dioxide.

4. In the process of respiratory gas exchange, the movement of _____ occurs by diffusion.
   a. respiratory gases across the respiratory surfaces of the lung
   b. respiratory gases between the tissues of the body and the circulatory system
   c. atmospheric gas into and out of the lungs
   d. respiratory gases within the blood of the circulatory system
   e. (a) and (b)
   f. (c) and (d)

5. Which of the following represents an evolutionary adaptation that terrestrial animals have to aid in improving the efficiency of their respiratory systems?
   a. keep the respiratory exchange surfaces outside the body to expose them to the oxygen in the air
   b. moisten the respiratory exchange surface only when necessary
   c. thicken the membrane lining the body surface to reduce water loss and promote the diffusion of gases through the skin
   d. development of complex internal lungs with massive amounts of surface area for gas exchange

6. An advantage of gas exchange in aquatic habitats, as compared to terrestrial, is that _____.
   a. water contains more dissolved oxygen than air
   b. gills are more protected from environmental damage
   c. keeping respiratory membranes moist is easy
   d. body heat is more easily maintained

7. Which of the following makes up the gas-exchange portion of the human respiratory system?
   a. larynx
   b. alveoli
   c. pharynx
   d. bronchi

8. A branch of the _____ brings deoxygenated blood to an alveolus.
   a. pulmonary artery
   b. pulmonary vein
   c. aorta
   d. inferior vena cava

9. Which of the following represents the role of the epiglottis in the human respiratory system?
   a. It blocks the esophagus during breathing to prevent air from going into the stomach.
   b. It contracts to force air out of the trachea.
   c. It blocks the larynx during swallowing to prevent food from entering the lungs.
   d. It acts as a gas-exchange surface for the exchange of carbon dioxide and oxygen.

10. Which structure is shared by the digestive and respiratory systems?
    a. esophagus
    b. alveolus
    c. larynx
    d. pharynx

11. Which part of the respiratory system is responsible for producing the voice?
    a. bronchioles
    b. larynx
    c. pharynx
    d. alveolus

12. Which of the following structures represents the smallest tubes within the conducting portion of the respiratory system?
    a. bronchioles
    b. bronchi
    c. trachea
    d. larynx

13. The major criterion that determines rate of breathing is _____.
    a. blood oxygen level
    b. blood carbon dioxide level
    c. nitrogen in the atmosphere
    d. blood carbon monoxide level

14. When you perform strenuous exercise, _____.
    a. the diaphragm and rib muscles will contract to increase chest volume
    b. excess carbon monoxide will stimulate the respiratory center of the brain so the rate of breathing increases
    c. oxygen will diffuse into the capillaries around the alveoli in addition to diffusing through the walls of the bronchi
    d. additional oxygen and carbon dioxide are exchanged through the skin to support the increase in metabolism

15. Carbon monoxide poisoning kills by _____.
    a. preventing the production of neurotransmitters in brain cells
    b. binding to oxygen, thereby preventing it from getting to hemoglobin
    c. depriving cells of oxygen by competing with it for hemoglobin
    d. enzymatically destroying the cells of the body

QUIZ 2

1. How would you BEST define the role of oxygen and carbon dioxide in aerobic cellular respiration?
   a. Carbon dioxide is required and oxygen is produced during aerobic cellular respiration.
   b. Oxygen is required and carbon dioxide is produced during aerobic cellular respiration.
   c. Both are required as building blocks for producing ATP.
   d. Neither is required for aerobic cellular respiration.

2. The role of bulk flow in respiration is to _____.
   a. exchange respiratory gases between inhaled air and blood
   b. exchange respiratory gases between blood and the body tissues
   c. transport gases carried in the blood and lung air

3. Oxygen diffuses from the blood into the tissues because
   a. the tissues have depleted their oxygen, and the oxygen in the blood is at a higher concentration than in the tissues
   b. tissues produce carbon dioxide, which attracts oxygen
   c. tissues produce oxygen and the oxygen in the blood wants to mix with the oxygen in the tissues
   d. oxygen is charged and is attracted to oppositely charged molecules in the tissues

4. Which features that facilitate diffusion of respiratory gases are shared by all animals with some type of respiratory system?
   a. The surface across which the exchange of respiratory gases is made is exceedingly thin.
   b. The respiratory surface is always moist.
   c. The respiratory surface is very big.
   d. The respiratory surface is an internal lung.
   e. All of the above are correct.
   f. (a), (b), and (c) are correct.

5. Which of the following animals use tracheae for respiration?
   a. mollusks
   b. snails

c. insects

d. fish

6. The "gas-exchange portion" of the respiratory system is the part of the anatomy in which
   a. the blood mixes together with air.
   b. gases in the air mix together.
   c. the only function is to carry air.
   d. respiratory gases are exchanged between the inhaled air within the alveoli and the blood flowing through the pulmonary capillaries.

7. Take a huge breath while sitting up straight in your chair, and then do this again while slumped over. Compare the air volume entering the lungs in each situation. Why is there a difference?
   a. The chest and rib cage cannot be increased in size effectively when you are slumped over in a chair.
   b. You cannot contract your chest muscles when slumped in a chair.
   c. There is less air moved into the lungs when you are slumped because the tissues need less oxygen for cellular respiration.
   d. The rib cage stays locked in an upward and outward orientation.

8. Compare the lungs to other respiratory structures such as gills and skin. What is the major advantage of the lungs?
   a. Oxygen diffuses faster across lung respiratory membranes than gill respiratory membranes.
   b. The lungs have more surface area, with accompanying greater respiratory membrane for gas exchange.
   c. Lungs are easier to inflate than gills.
   d. The alveoli of lungs are more protected than other respiratory membranes.

9. Sleep apnea is a cessation of breathing that can result in lack of sleep, irritability, lack of concentration, tiredness, and, in severe cases, heart attacks. The apnea is prompted either by a failure of the brain to communicate with the respiratory muscles or an obstructed airway associated with the tongue or muscles of the pharynx or larynx. The more efficient treatment, although one that many people prefer not to use, is CPAP—continuous positive airway pressure. Fitting over the nose, a mask is attached to a tube that is connected to a portable low-pressure generator. How is this breathing different from normal breathing?
   a. There is no difference between the breathing mechanics. The generator produces negative pressure breathing, causing the air to be sucked into the lung in a similar way to normal breathing.
   b. The generator increases the air pressure so that the air is actually forced into the lungs. This is the opposite of what happens in a normal breath.

c. The generator sucks the air out of the lungs actively, whereas in a normal expiration it is passively accomplished.

10. Through which sequence of structures would a volume of air have to travel from the mouth into the lungs?
    a. mouth—pharynx—bronchi—bronchioles—alveoli
    b. mouth—pharynx—bronchioles—bronchi—alveoli
    c. mouth—larynx—trachea—pharynx—alveoli
    d. none of the above

11. Which mechanism is used to transport both carbon dioxide and oxygen in the blood?
    a. hemoglobin
    b. blood plasma
    c. sodium bicarbonate ions
    d. all of the above

12. If humans did not have lungs, but instead used their thin and moist skin for gas exchange, why would they struggle to meet the oxygen demand of the tissues?
    a. The connection to the circulatory system would be lost.
    b. Temperature control would become very difficult.
    c. The external air does not contain a sufficient amount of oxygen for sufficient gas exchange.
    d. The skin would not provide a great enough surface area for gas exchange.

13. Why do many emphysema patients voluntarily contract muscles that help them increase their chest volume when they inhale?
    a. Emphysema patients have less respiratory surface area because of the damaged alveoli. To compensate, they must bring in larger volumes of air when they inhale.
    b. Emphysema patients have a higher metabolism. In order to meet the higher oxygen demand of the tissues, they must bring in larger volumes of air.
    c. The diaphragms of emphysema patients become paralyzed. The voluntary actions used to increase chest volume help compensate for this problem.
    d. The tumors associated with this disease obstruct the airways. The expansion of the chest cavity helps draw air past these obstructions.

14. A patient is rushed to the hospital unconscious. The doctor notes that the lips, extremities, and nail beds of this patient have turned blue. The patient's friend tells the doctor that he thinks it is carbon monoxide poisoning. Why should the doctor be skeptical of this diagnosis?
    a. Carbon monoxide would cause harm to humans only if it were at extremely high concentrations in the air we breathe.

b. Carbon monoxide breaks apart into carbon and oxygen. The oxygen can then bind to the hemoglobin in the red blood cells.

c. The blue color in the nail beds and lips means that the person has either lost a large amount of blood or blood is just not reaching these areas of the body.

d. When carbon monoxide binds to hemoglobin, it produces the bright red color seen when oxygen binds to this molecule.

15. When oxygen and carbon dioxide are carried in the blood, very little is carried in the plasma in a dissolved form. How does this help facilitate gas exchange at the lungs and tissues?

a. This helps pull oxygen into the blood from the air and carbon dioxide into the blood from the tissues.

b. The gases travel faster when they are bound to hemoglobin.

c. By competing for hemoglobin, the carbon dioxide can force the oxygen into the tissues. At the lungs, the oxygen forces the carbon dioxide off of this molecule so that it can be released into the air.

d. By removing the oxygen and carbon dioxide from solution in the plasma, a gradient for these gases can be maintained to favor diffusion.

ANSWER KEY

Quiz 1

1. c
2. e
3. b
4. e
5. d
6. c
7. d
8. a

9. c
10. d
11. b
12. a
13. b
14. a
15. c

Quiz 2

1. b
2. c
3. a
4. f
5. c
6. d
7. a
8. b

9. b
10. a
11. a
12. d
13. a
14. d
15. d

CHAPTER 34: NUTRITION AND DIGESTION

OUTLINE

Section 34.1 What Nutrients Do Animals Need?

- **Nutrients** provide the body with energy (measured as **calories**) and the raw materials needed to synthesize molecules necessary for life. There are six major forms of nutrients: (1) lipids, (2) carbohydrates, (3) proteins, (4) minerals, (5) vitamins, and (6) water.
- **Lipids** are used as an energy source, help form cell membranes, and are used in the synthesis of hormones and bile. Animals can store energy in the form of **fat**. **Essential fatty acids** cannot be synthesized by humans for the production of specific lipids and must be ingested.
- **Carbohydrates** are used as a source of quick energy but may be stored as starch or glycogen. The polysaccharide cellulose is used as the primary structural component in plants.
- **Proteins** play many structural and functional roles in the body and are composed of amino acids. **Essential amino acids** cannot be synthesized by humans and must be ingested.
- **Minerals** are elements that play an essential role in animal nutrition (**Table 34-2**).
- **Vitamins** are organic compounds that are required for normal cell function and control many metabolic reactions throughout the body (**Table 34-3**). Vitamins can be grouped as **water-soluble vitamins** and **fat-soluble vitamins**.
- **Water** is an important medium for internal transport and chemical reactions. Two-thirds of the human body is water.

Section 34.2 How Is Digestion Accomplished?

- Digestion acts to break down food into absorbable forms. Animal digestive systems are organ systems that help optimize digestive processes. This is accomplished through **ingestion**, **mechanical breakdown**, **chemical breakdown**, **absorption**, and **elimination**.
- Sponges are the only animals that digest their food by **intracellular digestion**, through the use of **collar cells**, **food vacuoles**, and **lysosomes** (**Figure 34-4**). Larger organisms digest their food by **extracellular digestion**.
- The simplest digestive systems (e.g., those found in cnidarians) consist of a sac with one opening (**gastrovascular cavity**) in which all extracellular digestive processes take place (**Figure 34-5**).
- Active animals that must eat frequently digest their food using a tubular digestive tract, composed of a one-way tube with openings at either end. This arrangement allows for the consumption of multiple meals and digestive specialization. Examples of digestive specializations include ruminant cellulose digestion (**Figure 34-6**), variations in small intestine length and in tooth forms (**Figure 34-7**), and bird gizzards (**Figure 34-8**).

Section 34.3 How Do Humans Digest Food?

- Human digestive systems are well adapted for processing a wide variety of foods (**Figure 34-9**). This is accomplished, in part, by the actions of powerful digestive secretions (**Table 34-4**).
- Mechanical and chemical breakdown of food begins in the mouth, through of the chewing action of teeth and the secretion of **amylase** (which begins breaking down starch into sugar) in saliva.
- Swallowing forces food from the mouth into the **esophagus** (**Figure 34-10**), which then transports it to the **stomach** by **peristalsis** (**Figure 34-11**). The stomach is an expansible muscular sac that stores and churns food, continuing its mechanical breakdown (**Figure 34-12**). The stomach also continues the chemical breakdown of food through the secretion of **proteases** and acidic stomach juices that start the protein digestion process.
- Most chemical digestion occurs in the small intestine, through the action of **bile** (which emulsifies fat droplets), **pancreatic juices** (which neutralize acidic chime and continue carbohydrate, lipid, and protein breakdown), and cells of the intestinal wall (which contain peptidases and disaccharidases). These processes are summarized in **Figure 34-13**. Most **absorption** takes place in the small intestine through the use of **villi** and **microvilli** in the intestinal lining (**Figure 34-14**). Most nutrients are absorbed into villus capillaries, but glycerol and fatty acids are absorbed by lymphatic capillaries (**lacteals**).
- The **large intestine** is responsible for water absorption and the formation of semisolid **feces**, which are composed of indigestible wastes, bacteria, and leftover nutrients.

- Stimuli from food (taste, smell, and visual appeal) trigger nervous system responses that prepare the digestive system for food digestion (**Figure 34-15**). The hormone **gastrin** is involved in the regulation of stomach acidity (**Figure 34-16**), the hormones **secretin** and **cholecystokinin** stimulate the release of bile and pancreatic juices into the small intestine, and the hormone **gastric inhibitory peptide** stimulates insulin release by the pancreas and inhibits stomach acid production and peristalsis.

LEARNING OBJECTIVES

Section 34.1 What Nutrients Do Animals Need?

- Describe the sources and metabolic uses for carbohydrates, proteins, minerals, vitamins, and lipids.
- Discuss the effects of vitamin deficiencies.
- Discuss the many uses of water in the body.

Section 34.2 How Is Digestion Accomplished?

- Describe the five steps of digestion.
- Describe the digestive systems of sponges, cnidarians, and insects.
- Explain how rumination allows animals to obtain nutrients from grasses.
- Discuss anatomical adaptations of animals that relate to their diet.

Section 34.3 How Do Humans Digest Food?

- Explain the process of human digestion from initial breakdown in the mouth to absorption in the small intestine.
- Explain how food is moved through the digestive system.
- Discuss the roles of the liver, gallbladder, and pancreas in human digestion.
- Discuss the specialized structures of the small intestine and how they relate to function.
- Describe the function of the large intestine.
- Describe the hormonal controls over the digestive system.

QUIZ 1

1. The major nutrient group used for quick energy production is _____.
   a. vitamins
   b. proteins
   c. lipids
   d. carbohydrates
   e. electrolytes

2. Why can vitamins A, D, E, and K be toxic, but C and B cannot be toxic?
   a. Fat-soluble vitamins such as C and B are stored in the body.
   b. Vitamins B and C do not affect brain chemistry.
   c. A, D, E, and K are used faster than C and B.
   d. Humans get more A, D, E, and K than C and B in their diets.

3. Essential amino acids _____.
   a. are the 10 or so required to make proteins
   b. can be synthesized in the body given the right fatty acid building blocks
   c. are not synthesized in the body; they must be taken in with food

4. Which of the following explains why the average herbivore has a longer intestine than the average carnivore?
   a. Carnivores do not have to digest their food mechanically.
   b. Carnivores absorb many of their nutrients from the stomach.
   c. The cell walls in plants are difficult to digest. For herbivores, the longer intestine provides more opportunity to extract nutrients from plant material.
   d. Carnivores produce the enzyme cellulase. This enzyme helps in the digestion of cellulose found in the cell walls of plant cells. This efficient digestion of plant material means that carnivores can have shorter intestines.

5. Digestion in cnidaria, as in hydra, occurs in a
   _____.
   a. crop
   b. coelom
   c. digestive tract

d. gastrovascular cavity
e. pharynx

6. Collar cells are found in _____ .
  a. sponges
  b. fish
  c. insects
  d. tapeworms
  e. hydra

7. The role of the liver in the digestion of food is to produce _____ .
  a. lipase for the digestion of fat
  b. bile for the emulsification of fat
  c. pepsin for the digestion of protein
  d. bile for the digestion of carbohydrates

8. The hormone responsible for stimulating the secretion of hydrochloric acid by stomach cells is
  a. pepsin.
  b. gastrin.
  c. cholecystokinin.
  d. insulin.
  e. secretin.

9. A sudden increase in the amount of secretin circulating in the blood is an indication that food has recently been introduced into the
  a. mouth.
  b. pharynx.
  c. stomach.
  d. small intestine.
  e. large intestine.

10. Which organs produce secretions involved in the chemical breakdown of carbohydrates?
  a. salivary glands
  b. stomach
  c. large intestine
  d. pancreas
  e. (a) and (d)
  f. (b) and (c)

11. The role of amylase in digestion is the
  a. chemical breakdown of disaccharides into monosaccharides.
  b. chemical breakdown of polysaccharides into amino acids.
  c. mechanical breakdown of carbohydrates into monosaccharides.
  d. chemical breakdown of polysaccharides into disaccharides.

12. Which organ produces the enzymes that are involved in the breakdown of lipids (fats)?
  a. pancreas
  b. liver
  c. salivary glands
  d. large intestine

13. Which of the following represents the role of the gallbladder in digestion?
  a. It stores food and then releases it slowly to the small intestine.
  b. It stores bile and releases it into the small intestine to emulsify fat.
  c. It stores pancreatic juices and releases them into the small intestine.
  d. It produces enzymes that are involved in the chemical breakdown of proteins.

14. Which of the following is chemically broken down by pancreatic secretions?
  a. carbohydrates
  b. proteins
  c. lipids
  d. (a) and (b)
  e. (b) and (c)
  f. all of the above

15. The secretion from the stomach is involved in the chemical breakdown of
  a. carbohydrates.
  b. proteins.
  c. lipids.

QUIZ 2

1. A (an) _____ raises the temperature of a gram of water by 1 degree Celsius.
  a. Calorie
  b. calorie
  c. fat gram
  d. ATP

2. "Water-soluble compound that works primarily as an enzyme helper" would be a good definition of
  a. vitamin B.
  b. vitamin A.
  c. vitamin E.
  d. vitamin D.

3. Food particles are broken down into smaller pieces to make enzyme activity more efficient. This would best be described as
  a. chemical breakdown.
  b. absorption.
  c. elimination.
  d. mechanical breakdown.

4. Which is the correct sequence of events that all digestive systems must follow in the processing of foodstuffs?
  a. ingestion—chemical breakdown—mechanical breakdown—absorption—elimination

  b. ingestion—mechanical breakdown—chemical breakdown—absorption—elimination

  c. ingestion—absorption—mechanical breakdown—chemical breakdown—elimination

  d. ingestion—absorption—chemical breakdown—mechanical breakdown—elimination

5. An earthworm's digestive anatomy is designed so that this animal can feed more frequently than an animal such as a sea anemone. Which of the following explains this relationship?

  a. A sea anemone does not produce any digestive enzymes.

  b. A sea anemone cannot ingest its food.

  c. An earthworm has a highly efficient ruminant stomach.

  d. An earthworm has a tubular digestive tract, while a sea anemone has a digestive system with only one opening.

6. Chyme is _____.

  a. a protein-digesting enzyme

  b. the thickened food material after digestion in the stomach

  c. enzymatic juice secreted by the small intestine

  d. the fatty material absorbed into lacteals, also called *chylomicrons*

  e. an enzyme that breaks down complex carbohydrates into disaccharides

7. The cause of ulcers is _____.

  a. *Helicobacter pylori*

  b. stress

  c. alcoholism

  d. allergies

  e. smoking

8. The function of bile is the _____.

  a. facilitation of sugar and amino acid absorption

  b. mechanical breakdown of fats for better breakdown with lipase, a lipid-digestive enzyme

  c. production of bile pigments

  d. production of glucose

  e. maintenance of blood glucose level

9. The nasal cavities and oral cavity merge into the pharynx. What keeps food and drink from going into the nasal cavity rather than down the esophagus?

  a. The tongue blocks the upper pharynx leading into the nasal cavities.

  b. The soft palate moves up against the opening to the nasal cavities, blocking the entryway.

  c. Breathing is temporarily stopped so that the nasal cavities can close when swallowing.

  d. The esophagus constricts at the upper end of the pharynx so the entry to the nasal cavities is blocked.

10. Weight loss, anemia, and fatigue are symptoms of

  a. gallstones. The hardened bile in the gallbladder affects lipid metabolism, causing weight loss and fatigue. The anemia is due to a lack of bile pigments available to produce red blood cells.

  b. peptic ulcer. The bleeding ulcer causes loss of blood, resulting in anemia and fatigue. The weight loss is caused by the inability of the stomach to absorb nutrients.

  c. intestinal cancer. There is less surface area for nutrient absorption, hence the weight loss. The lack of nutrients also accounts for the inability to make the hemoglobin and red cells, leading to anemia, and for the lack of oxygen-carrying capacity, fatigue.

  d. colitis. The inflammation of the colon makes it impossible for nutrients to be properly absorbed, hence the weight loss and fatigue. The inflammation also disrupts the absorption of various vitamins required to make hemoglobin and red blood cells and reduces oxygen-carrying capacity.

11. Food entering the _____ triggers the swallowing reflex.

  a. mouth

  b. larynx

  c. pharynx

  d. trachea

12. Which of the following is NOT a function of the stomach?

  a. Protein digestion begins due to the actions of pepsin.

  b. Rhythmic contractions known as *peristaltic movements* move food into the stomach and aid in the mixing of the partially digested contents and the gastric secretions (chyme) of the stomach.

  c. It produces a thick mucus secretion along the stomach lining that protects the stomach from being digested by its own secretions (e.g., pepsin).

  d. It is one of the primary sites for the absorption of digested protein fragments.

13. Why does the small intestine absorb more nutrients than any other region of the digestive tract?

  a. It is the site of segmentation movements.

  b. The exposed surface area approaches that of a tennis court in size.

  c. It contains a concentrated supply of capillaries that lie in intimate contact with this region of the digestive tract.

  d. All of the above are correct.

14. Starch (amylose) digestion is complex and involves several different parts of the digestive anatomy. Which of the following represents the places in the human body where starch is chemically digested?

  a. mouth and liver

  b. salivary gland and pancreas

    c. mouth and large intestine
    d. mouth and small intestine

15. Which of the following is the active form of an enzyme that chemically digests proteins in the stomach?

    a. pepsin
    b. pepsinogen
    c. lipase
    d. secretin

ANSWER KEY

Quiz 1

1. d		**9.** d
2. a		**10.** e
3. c		**11.** d
4. c		**12.** a
5. d		**13.** b
6. a		**14.** f
7. b		**15.** b
8. b		

Quiz 2

1. b		**9.** b
2. a		**10.** c
3. d		**11.** c
4. b		**12.** d
5. d		**13.** d
6. b		**14.** d
7. a		**15.** a
8. b		

CHAPTER 35: THE URINARY SYSTEMS

OUTLINE

Section 35.1 What Are the Basic Functions of Urinary Systems?

- All urinary (excretory) systems perform three functions: (1) filter blood or interstitial fluid, thus removing water and small dissolved molecules; (2) selectively reabsorb nutrients from filtrate; and (3) excrete remaining water and dissolved wastes from the body.

Section 35.2 What Are Some Examples of Invertebrate Excretory Systems?

- Flatworm excretory systems consist of branching **protonephridia** that carry fluids and wastes to excretory pores that lead out of the body (**Figure 35-1a**). Flatworms also have a large body surface area through which most cellular wastes diffuse out.
- Insects use **Malphigian tubules** that filter hemocoel blood and release wastes, salts, and water into the filtrate that is deposited into the intestine. The intestine and rectum reabsorb salts and water, producing concentrated urine that is released with feces (**Figure 35-1b**).
- Earthworms, mollusks, and several other invertebrates use **nephridia** that filter the interstitial fluid that fills the body cavity. Salts and nutrients are reabsorbed, leaving water and wastes that are excreted through the **nephridiopore** opening in the body wall (**Figure 35-1c**).

Section 35.3 What Are the Functions of Vertebrate Urinary Systems?

- Vertebrate urinary systems regulate blood ion, water, and nutrient levels; maintain blood pH; secrete hormones; and eliminate cellular wastes.
- One important function of excretory systems is the elimination of nitrogenous wastes that are formed from protein breakdown (**Figure 35-2**). Nitrogenous wastes can be eliminated as ammonia, urea, or uric acid.

Section 35.4 What Are the Structures and Functions of the Human Urinary Systems?

- The human urinary system is composed of **kidneys, ureters**, a **bladder**, and a **urethra** (**Figure 35-3**).
- Each kidney is composed of more than 1 million urine-forming **nephrons** that are located in the **renal cortex** and **renal medulla** (**Figure 35-4**). Each nephron is composed of a **glomerulus, Bowman's capsule**, and **tubule** (**Figure 35-5**). Fluid within the nephron tubule is received by a **collecting duct**, which deposits urine in the **renal pelvis** of the kidney. The renal pelvis then funnels urine into the **ureter** and then into the **bladder** where it is collected. Distension of the muscular bladder triggers urination, which causes urine to flow through the **urethra** on its way out of the body.
- Blood is filtered under pressure through capillaries in the glomerulus, which deposits water and dissolved substances as filtrate in the Bowman's capsule (**Figure 35-6**). Filtrate in the Bowman's capsule flows into the nephron tubule, where the active transport of nutrients and osmotic movement of water into the capillary blood takes place (**tubular reabsorption**). Some wastes are actively transported into the distal tubule filtrate from capillary blood (**tubular secretion**).
- The nephron **loop of Henle** generates a salt gradient in the kidney interstitial fluid surrounding the nephron. This causes the osmotic movement of water from the filtrate into the interstitial fluid of the kidney, thus concentrating urine. If **antidiuretic hormone** (**ADH**) is present, additional water leaves the urine as it flows through the collecting duct (**Figure E35-1**).

Section 35.5 How Do Mammalian Kidneys Help Maintain Homeostasis?

- Mammalian kidneys regulate the water content of the blood by the release of **ADH**, which increases the permeability of the collecting ducts to water. ADH release is stimulated by water loss and dehydration (**Figure 35-7**).
- The kidneys regulate blood pressure by releasing the hormone **renin** into the bloodstream. When blood pressure is low, renin is released by the kidney and catalyzes the formation of the hormone **angiotensin**. Angiotensin causes arterioles to constrict, elevating blood pressure.

- The kidneys regulate blood oxygen levels by releasing the hormone **erythropoietin** when blood oxygen levels are low, which stimulates bone marrow to produce red blood cells.
- The kidneys regulate the concentration of dissolved substances in the blood by controlling tubular secretion and reabsorption.
- Loops of Henle determine the urine-concentrating capacity of mammals. Mammals that live in water-abundant habitats (e.g., beavers) have short loops of Henle, resulting in the production of dilute urine. Mammals that live in desert environments have long loops of Henle, resulting in the production of concentrated urine.

LEARNING OBJECTIVES

Section 35.1 What Are the Basic Functions of Urinary Systems?

- Discuss the two main functions of the urinary system.

Section 35.2 What Are Some Examples of Invertebrate Excretory Systems?

- Describe the excretory system of a flatworm.
- Describe the function of malpighian tubes.

Section 35.3 What Are the Functions of Vertebrate Urinary Systems?

- Discuss the role of vertebrate kidneys in maintaining homeostasis.
- Discuss the difference between fish and terrestrial animals in their excretion of ammonia.
- Discuss how birds and amphibians conserve water while eliminating ammonia waste products.

Section 35.4 What Are the Structures and Functions of the Human Urinary Systems?

- Describe the anatomy of the human kidney.
- Explain how the structure of the Bowman's capsule facilitates the filtration of blood.
- Describe the process of and necessity for tubular reabsorption.
- Discuss which substances are actively transported during tubular secretion.

Section 35.5 How Do Mammalian Kidneys Help Maintain Homeostasis?

- Describe the effect of antidiuretic hormone secretion on the volume and concentration of urine produced.
- Discuss the trigger for renin release and its effect on kidney function.
- Discuss the role of kidney failure on anemia.
- Discuss the osmolarity issues faced by freshwater animals and saltwater animals.
- Describe the relationship between the length of the Loop of Henle and water conservation.

QUIZ 1

1. Freshwater trout are urinating all of the time and producing many times their own blood volume over the course of a 24-hour day, while the desert-dwelling kangaroo rat might excrete only a few milliliters over the same time period. The excretory mechanisms in both animals are attempting to
   a. regulate the water volume within the body.
   b. maintain a constant level of acidity within the body.
   c. govern the levels of the same ions (e.g., Na^+ and K^+).
   d. all of the above.
   e. none of the above.

2. Flame cells are the specialized excretory structures found in _____.
   a. flatworms
   b. earthworms
   c. protozoa
   d. insects

3. For affected species, what physiological advantage does the ability to excrete uric acid have over the ability to excrete urea?
   a. Fewer electrolytes (ions) are lost.
   b. Excretion of uric acid enables a greater volume of urine to be produced.
   c. More electrolytes are lost.

d. Very little water loss occurs as a result of excreting uric acid.

e. (a) and (d) are correct.

4. Which of the following does NOT become part of the filtrate formed by the glomerulus?
   a. plasma proteins
   b. water
   c. salt
   d. glucose

5. Which of the following must be reclaimed from the filtrate during the process of reabsorption?
   a. hydrogen ions
   b. glucose
   c. drugs
   d. toxins

6. Which portion of the loop of Henle is highly permeable to water?
   a. descending portion
   b. thick portion of the ascending loop
   c. thin portion of the ascending loop

7. Which portion of the loop of Henle actively transports salt into the kidney medulla?
   a. descending portion
   b. thick portion of the ascending loop
   c. thin portion of the ascending loop

8. Which of the following is secreted from blood into the filtrate at the distal tubule?
   a. glucose
   b. drugs
   c. amino acids
   d. salt

9. Diuretics are substances that cause the production of dilute urine. Which of the following descriptions represents a possible mechanism of action for a diuretic?
   a. increases the release of ADH from the pituitary of the brain
   b. promotes salt reabsorption
   c. It has a direct effect on the collecting duct by increasing the permeability of this part of the nephron
   d. blocks the effects of ADH

10. What effect would constriction of the incoming arteriole have on the filtration pressure in the glomerulus?
    a. The glomerular pressure and the amount of filtrate produced would increase as a result of the constricted afferent arteriole.
    b. The constricted afferent arteriole would not change the already smaller size of the outgoing arteriole, thus the glomerular pressure and amount of filtrate would not change.
    c. The glomerular pressure would decrease and the amount of filtrate produced would decrease as a result of the constricted afferent arteriole.

d. The glomerular pressure would decrease and the amount of filtrate produced would increase as a result of the constricted afferent arteriole.

11. After leaving the proximal tubule, the filtrate enters
    a. Bowman's capsule.
    b. the loop of Henle.
    c. the distal tubule.
    d. the collecting duct.

12. In urine formation, the role of ADH (antidiuretic hormone) is to cause
    a. the collecting duct to become highly permeable to water.
    b. glomerular capillaries to dilate, which increases blood flow into the glomerulus.
    c. the collecting duct to become impermeable to water.
    d. the collecting duct to constrict to reduce the flow of filtrate out of a nephron.

13. A diuretic is any substance that causes an increase in urine output. Alcohol acts as a diuretic because it
    a. causes the release of endorphins from the brain. The endorphins help stimulate ADH release.
    b. causes the release of endorphins from the brain. The endorphins inhibit ADH release.
    c. stimulates the production of angiotensin in the blood.
    d. stimulates the release of erythropoietin from the kidney.

14. Which of the following could be a sign of a failing kidney?
    a. a drop in the number of red blood cells
    b. dramatic changes in systemic blood pressure
    c. water retention and swelling in the tissues
    d. all of the above

15. A very common type of drug used to control high blood pressure is an ACE inhibitor. This drug inhibits an enzyme involved in the production of a hormone called angiotensin. Which of the following explains why blood pressure would drop if you were taking this drug?
    a. Angiotensin production drops and more water is retained in the blood. The increase in blood volume helps lower blood pressure.
    b. Angiotensin production drops and more water is removed from the blood. The decrease in blood volume helps lower blood pressure.
    c. There is less angiotensin traveling to the kidneys. This leads to a constriction of the blood vessels leading to the kidneys and a lower filtration rate.
    d. There is less angiotensin traveling to the kidneys. This leads to a constriction of the blood vessels leading to the kidneys and a higher pressure in the glomerulus.

QUIZ 2

1. What similar functions do the protonephridia of flatworms and the nephridia of roundworms have in common?
   a. Almost none; the species are too different.
   b. Both act to filter out wastes for subsequent excretion.
   c. Both act to filter out nutrients to retain them inside the body.
   d. (b) and (c) are correct.
   e. None of the above is correct.

2. Which organ converts ammonia to urea?
   a. liver
   b. pancreas
   c. gallbladder
   d. kidney

3. When a person takes a drug, the drug will eventually be eliminated from the body. One of the primary mechanisms for this removal is tubular secretion. Which of the following would produce the greatest reduction in the ability of our kidneys to remove drugs?
   a. a reduction in the salt gradient within the renal medulla
   b. damage to Bowman's capsule
   c. damage to the wall of the urinary bladder
   d. damage to the distal tubule of the nephron

4. What determines the ability of a mammal to concentrate its urine?
   a. number of nephrons
   b. length of the tubules
   c. length of the collecting duct
   d. size of the glomerulus
   e. length of the loop of Henle

5. The muscular tube that transports urine formed by a kidney to the bladder is called a
   a. ureter.
   b. urethra.
   c. vena cava.
   d. glomerulus.

6. What part of the kidney contains the nephrons?
   a. renal pelvis
   b. renal vein
   c. renal cortex
   d. urinary bladder

7. What happens to a substance that is neither filtered nor reabsorbed but is secreted?
   a. The body retains it.
   b. It is eliminated from the body.
   c. Its concentration in the blood gradually decreases.
   d. Its concentration in the blood gradually increases.
   e. (b) and (c) are correct.
   f. (a) and (d) are correct.

8. Blood is filtered between _____.
   a. the proximal tubule and the distal tubule
   b. the descending limb of the loop of Henle and the ascending limb of the loop of Henle
   c. the collecting duct and Bowman's capsule
   d. Bowman's capsule and the capillary bed known as the *glomerulus*

9. In the human urinary system, the kidneys _____ urine.
   a. store
   b. transport
   c. produce
   d. excrete

10. Tubular reabsorption is the _____.
    a. process by which wastes and excess substances that were not initially filtered out into Bowman's capsule are removed from the blood
    b. concentration of the filtrate resulting from the concentration of salts and urea
    c. passage of blood cells and proteins from the glomerulus
    d. process by which cells of the proximal tubule remove water and nutrients and pass them back into the blood

11. What initiates glomerular pressure filtration?
    a. the permeability of the glomerular walls
    b. the inability of large proteins to pass through the capillary walls
    c. diameter differences between the incoming and outgoing arterioles
    d. osmotic pressure differences between the glomerulus and Bowman's capsule

12. The difference between the blood plasma that exits the kidney by way of the renal vein and the plasma that enters the kidney by way of the renal artery is that the exiting plasma
    a. contains fewer dissolved nutrients.
    b. contains more dissolved nutrients.
    c. contains fewer dissolved wastes.
    d. contains more dissolved wastes.

13. The hormone ADH _____.
    a. controls the concentration of urine
    b. controls the amount of blood that enters the kidney
    c. controls the reabsorption of nutrients
    d. determines the pH of the blood

14. Angiotensin is a hormone that _____.
    a. causes arterioles to constrict
    b. regulates the amount of water reabsorbed from the urine
    c. catalyzes the formation of angiotensin
    d. stimulates the production of red blood cells

15. Under which of the following conditions would you expect the production and secretion of the hormone ADH to increase?
    a. You have been trapped, unharmed, for a day under the rubble that used to be your house before a large storm destroyed it.
    b. You have just suffered your first loss of the 4th of July softball tournament, and now your team will need to play six games in one day with only two 1-hour breaks to reach the championship game.
    c. A plumber came out early in the morning and shut off the water supply to your new house so that a new water main could be put in. All that you've been doing is sitting around the house in the shade waiting for the water to come back on and for the delivery of your new dishwasher and refrigerator. The water to the house isn't turned on until 7 P.M. that evening.
    d. All of the above are correct.
    e. None of the above is correct.

ANSWER KEY

Quiz 1

1. d		**9.** d	
2. a		**10.** c	
3. e		**11.** b	
4. a		**12.** a	
5. b		**13.** b	
6. a		**14.** d	
7. b		**15.** b	
8. b			

Quiz 2

1. d		**9.** c	
2. a		**10.** d	
3. d		**11.** c	
4. e		**12.** c	
5. a		**13.** a	
6. c		**14.** a	
7. e		**15.** d	
8. d			

CHAPTER 36: IMMUNITY: DEFENSES AGAINST DISEASE

OUTLINE

Section 36.1 What Are the Basic Mechanisms of Defense Against Disease?

- Vertebrates have three major lines of defense against disease: (1) **nonspecific external barriers**, (2) **nonspecific internal defenses**, and (3) **specific internal defenses (Figure 36-1)**.
- Invertebrates have two major lines of defense against disease: (1) **nonspecific external barriers** and (2) **nonspecific internal barriers**.

Section 36.2 How Do Nonspecific Defenses Function?

- **Nonspecific external defenses**, which include the **skin** and **mucous membranes**, keep pathogens from entering the body. The **skin** is a physical barrier for pathogens, while its secretions prevent bacterial growth. **Mucous membranes** secrete mucus and antimicrobial substances that trap and destroy pathogens. Respiratory mucous membranes contain cilia that sweep microbes trapped in mucus to the nose and mouth where they can be removed from the body (**Figure 36-3**).
- **Nonspecific internal defenses** attack pathogens that penetrate nonspecific external defenses. **Phagocytes** engulf pathogens, while **natural killer cells** secrete proteins that kill virally infected or cancerous cells. Tissue damage stimulates the **inflammatory response (Figure 36-5)** in which secreted chemicals (including **histamine**) stimulate phagocyte activity and increase blood flow and capillary permeability. Clotting eventually seals off the wound site, preventing additional microbe entry. **Fever** occurs when endogenous pyrogens are released by white blood cells in response to infection. Elevated temperature inhibits bacterial growth, accelerates immune cell activity, and causes the secretion of virus-inhibiting **interferons** by infected cells.

Section 36.3 What Are the Key Characteristics of the Immune Response?

- The immune response involves the activity of two types of lymphocytes: **B cells** and **T cells (Table 36-1)**. B cells form plasma cells that secrete **antibodies** into the bloodstream as part of **humoral immunity**. **Cytotoxic T cells** can destroy some microbes, cancer cells, and virally infected cells as part of **cell-mediated immunity**. Helper T cells stimulate both **humoral immunity** and **cell-mediated immunity**.
- The immune response has three steps: (1) **recognition**, (2) **attack**, and (3) **memory**.
- In the **recognition** step, antibodies (made by B cells or plasma cells) or T cell receptors recognize foreign **antigens** and trigger the immune response. Antibodies are Y-shaped proteins that bind to and help destroy antigens (**Figures 36-6** and **36-7**). Each B cell produces only one type of antibody. Each antibody binds to only one or a few types of antigens.
- In the **attack** step, antigens from a pathogen bind to B or T cells with complementary membrane-bound antibodies or T cell receptors. In **humoral immunity**, B cells bound to a specific antigen undergo **clonal selection**, resulting in the production of **plasma cells** that produce massive amounts of identical antibodies complementary to the original antigen (**Figure 36-9**). In **cell-mediated immunity**, T cells (with proper receptors) bind to complementary antigens and divide rapidly. Activated **cytotoxic T cells** bind to antigens on microbes, cancer cells, or infected cells and kill the cells. Activated **helper T cells** chemically stimulate B cell and cytotoxic T cell responses. **Figure 36-11** summarizes this process.
- In the **memory** step, some of the B and T cells that are produced form long-lived **memory cells** that are activated if the same antigen reappears in the bloodstream. This produces a much faster and effective immune response (**Figure 36-12**).

Section 36.4 How Does Medical Care Augment the Immune Response?

- **Vaccinations** are injections of pathogenic antigens or weakened pathogenic microbes. This stimulates the development of memory cells that rapidly fight off future infections.

- **Antibiotics** are drugs that slow the growth and reproduction of bacteria, fungi, and protists. This allows the body more time to generate an immune response that will destroy these invaders.

Section 36.5 What Happens When the Immune System Malfunctions?

- **Allergies** occur when an immune response forms against a normally harmless substance. B cells respond to these substances (which are treated like antigens) by producing antibodies that bind to **mast cells**. Mast cells release **histamine** when exposed to these harmless antigens, which causes a local inflammatory response (**Figure 36-13**).
- **Autoimmune diseases** occur when the body forms an immune response against its own molecules.
- **Immune deficiency diseases** occur when the body cannot form an immune response strong enough to ward off pathogens.
- **AIDS** (acquired immune deficiency syndrome) occurs when human immunodeficiency viruses (**HIV**) destroy helper T cells, causing an individual to become extremely susceptible to infections.
- **Cancer** occurs when body cells multiply without control. These cells can be destroyed by natural killer cells or cytotoxic T cells, but if cells multiply too quickly for the immune system to keep up, a **tumor** develops.

LEARNING OBJECTIVES

Section 36.1 What Are the Basic Mechanisms of Defense Against Disease?

- Describe the nonspecific barriers to disease possessed by animals.
- Discuss the differences and similarities between vertebrate and invertebrate defense systems.

Section 36.2 How Do Nonspecific Defenses Function?

- Discuss how the skin provides protection from infection.
- Discuss the defense mechanisms found in mucous membranes.
- Describe the functions of neutrophils, macrophages, and natural killer cells.
- Explain the initiation and effect of the inflammatory response on invading organisms.
- Explain the initiation and effect of fever on invading organisms

Section 36.3 What Are the Key Characteristics of the Immune Response?

- Discuss the variety of molecules recognized as antigens.
- Describe the structure of an antibody.
- Explain how antibodies bind to their antigen.
- Discuss the differences between B cells and T cells.
- Discuss the identification of "self" as it relates to the immune system.
- Explain humoral immunity and how it provides long-term protection from reinfection by the pathogen.
- Describe the functions of the three types of T cells.
- Discuss the effects of HIV infection on the immune system.

Section 36.4 How Does Medical Care Augment the Immune Response?

- Explain how vaccines provide protection against a specific disease.
- Discuss the role of antibiotics in fighting bacterial infections.

Section 36.5 What Happens When the Immune System Malfunctions?

- Describe the immune system's role in allergic reactions.
- Discuss the various autoimmune disorders and their treatments.
- Discuss the role of natural killer and cytotoxic T cells in fighting cancer.
- Describe the three main forms of cancer treatment.

QUIZ 1

1. Most microorganisms ingested in food are destroyed by _____.
   a. stomach acid
   b. cilia
   c. saliva
   d. mucus
   e. interferon

2. If a microorganism is able to invade the internal tissues, bypassing physical and chemical barriers, it will likely be destroyed by _____.
   a. the skin
   b. the inflammatory response
   c. mucous membrane
   d. protein-digesting enzymes
   e. ciliary action

3. Nosocomial diseases are infectious diseases that are obtained in a hospital (or related) setting. The most common type of nosocomial infection is a urinary tract infection (UTI) associated with catheters. A catheter is a hollow tube apparatus that is inserted up the urethra of the urinary tract into the otherwise sterile urinary bladder. It is used to eliminate urine in people who have lost (either permanently or temporarily) the ability to void urine. Why are UTIs and catheters so often associated?
   a. The catheters are not sterile or in sterile packages. Bacteria on the catheters can ascend up the urethra to the bladder.
   b. Improper handling of the catheters contaminates them and they become conduits for bacteria from the outside to the inside, thereby bypassing the host defenses such as skin and mucous membranes.
   c. The catheters cause injury to the mucous membrane of the urethra, making the tissue susceptible to inflammation.
   d. The insertion of the catheter produces small breaks in blood vessels lining the urethral membrane, allowing bacteria from the blood to enter the urethra and eventually the urinary bladder.

4. The first line of defense against body cells infected by viruses is produced by
   a. antibodies.
   b. phagocytes.
   c. natural killer cells.
   d. histamines.
   e. natural antibiotics.

5. Your 5-year-old daughter has just come home from day care feeling poorly and not eating. The next day, she has vomiting, a rash, diarrhea, and some neurological symptoms. The diagnosis is bacterial meningitis, a contagious bacterial infection, usually not life threatening. Within a week, she is up and about again, apparently healthy. The little boy who lives next door to you gets this same disease 3 months later, and you become worried that your daughter will again get meningitis. They play together all the time, and he could easily transmit the bacterium to her during play. Is this a potential problem?
   a. Yes, your daughter is at risk for contracting meningitis again. Kids under the age of 10 years have immature immune systems that will not protect them. Maternal antibodies have long since dissipated in her body, leaving her exposed to all sorts of pathogenic microorganisms.
   b. Yes, your daughter is at risk for meningitis again. Bacteria do not elicit as strong an immune response as viruses do, and her immune system will not be able to contain the infection.
   c. No, your daughter is not at risk for contracting meningitis again, particularly since the time span between the first exposure and second exposure to this infectious agent is not very long. Upon "seeing" the bacteria again, her memory cells from the previous exposure will become activated quickly, again making large amounts of protective antibody.
   d. No, your daughter is not at risk for meningitis again. Once you have contacted an infectious agent and memory cells have been produced, that infectious agent will always be identified and eradicated before the infection can take place.

6. Which of the following is a limitation of the humoral response to microbial invasion?
   a. B cells are capable of attacking only microbial invaders that have taken over host cells.
   b. The humoral response is capable of effectively attacking invading microbes only before they enter a host cell.
   c. B cells have a limited ability to stop the immune response once the microbes have been neutralized.
   d. B cells indiscriminately devour invading microbes.

7. Can you have immunity to a microbe that you have never encountered?
   a. Yes, if you have encountered the same or a very similar antigen of another microbe.
   b. Yes. Your immune system has as many memory cells as it has antibodies.

179

c. No. You must encounter an entire microbe before you can become immune to it.

d. No. You cannot establish immunity against any microbes.

8. Why is the initial immune response to an antigen slower than and not as strong as the second exposure to the antigen?

a. The memory cells are not active during the first exposure.

b. There is a delay while the B-cell clones are selected, multiply, and differentiate.

c. The B cells involved in the initial exposure are slower than those involved in the second exposure.

d. The helper T cells involved in the initial exposure are slower than those involved in the second exposure.

9. Which of the following "selects" which B cells will become active during the immune response to an antigen?

a. T cells

b. other B cells

c. antibodies

d. the antigen

10. During differentiation of B cells, which type of cell becomes filled with rough endoplasmic reticulum to enable it to "crank out" huge numbers of antibodies?

a. plasma cells

b. macrophages

c. T helper cells

d. memory cells

11. Do antibiotics work on viruses? Why or why not?

a. Yes, antibiotics can disrupt the cellular membranes of bacteria, protists, and viruses.

b. Yes, antibiotics kill cells that use DNA to reproduce.

c. No, viruses must enter the body's cells to reproduce and are "protected" from antibiotics.

d. No, viruses use RNA to replicate; therefore, antibiotics do not affect them.

12. Vaccinations are good at providing immunity against _____.

a. bacteria

b. protists

c. viruses

d. (a) and (b)

e. (b) and (c)

f. all of the above

13. Since allergies are such a nuisance and allergens are not harmful to us, why do we even make allergy antibodies?

a. Allergens used to be harmful, but we have evolved resistances to them, so allergy antibodies are just evolutionary baggage.

b. Allergy antibodies help defend us against parasites that typically enter the body through the mouth, nose, and throat by increasing mucus secretions and coughing or sneezing to expel the parasites.

c. Allergy antibodies are just a mistake of the immune system and have no real use.

d. Allergens are harmful to us and allergy antibodies protect us from them.

14. AIDS and the infectious agent HIV are difficult to treat because HIV _____.

a. reproduces more quickly than any other virus known to humans

b. can mutate rapidly into drug-resistant forms

c. cannot be destroyed by antibiotics

d. spreads from organ to organ

15. Why is chemotherapy, but not radiation therapy, used to treat cancers that have spread throughout the body?

a. Radiating the entire body would cause too much healthy tissue damage.

b. Chemotherapy kills rapidly dividing cells regardless of location in the body.

c. Cancerous cells become resistant to radiation once they have begun to spread.

d. (a) and (b) are correct.

e. (b) and (c) are correct.

QUIZ 2

1. Fever

a. decreases interferon production.

b. decreases the concentration of iron in the blood.

c. decreases the activity of phagocytes.

d. increases the reproduction rate of invading bacteria.

e. does all of the above.

2. The role of histamine in the inflammatory response is to cause

a. the dilation of arterioles, leading to increased blood flow to the injured tissue.

b. larger numbers of white blood cells to be released from the bone marrow.

c. capillary permeability to increase as capillaries become "leaky."

d. (a) and (c).

3. An inflamed tissue turns red because of
   a. an increased movement of fluid from blood into tissues across "leaky" capillary walls.
   b. an increased number of white blood cells circulating in the blood.
   c. an increased blood flow to the injured tissues resulting from the dilation of arterioles in the injured area.
   d. the recruitment of macrophages to the injured tissue.

4. An inflamed tissue swells because of
   a. an increased movement of fluid from blood into tissues across "leaky" capillary walls.
   b. an increased number of white blood cells circulating in the blood.
   c. an increased blood flow to the injured tissues resulting from the dilation of arterioles in the injured area
   d. the recruitment of macrophages to the injured tissue.

5. An antigen is
   a. a substance produced by B cells to attack microbes.
   b. a receptor found on a B cell that is important for B-cell activation.
   c. a molecule released by T cells to activate B cells.
   d. a molecule found on the surface of invading microbes that is recognized by the immune system.

6. When a cell "differentiates," it is
   a. dividing.
   b. becoming more specialized in function.
   c. dying.
   d. becoming more generalized in function.

7. Which of the following cells is responsible for producing antibodies?
   a. red blood cell
   b. T cell
   c. B cell
   d. macrophage

8. Major histocompatibility complex (MHC) antigens are _____.
   a. antigenic determinants on microorganisms
   b. proteins produced by newly formed cancer cells
   c. a type of antibody molecule
   d. the binding sites of the HIV virus
   e. recognition markers on cells used for identification by one's own immune system

9. Which of the following statements regarding antibody is true? An antibody _____.
   a. has one binding site for antigens
   b. is found in a variety of body fluids and organ systems

   c. can be given to treat bacterial infections as an antibiotic
   d. is produced by platelets
   e. once produced, protects for the rest of one's life

10. In order for transplanted organs to be successfully accepted by the recipient's body, the donor and recipient must be matched and the recipient must be placed on medication, even for a long period after the transplant. Why are organ transplants such a problem?
    a. We do not yet have the technology to transplant organs successfully.
    b. Unless every donor and recipient are close relatives, their MHC proteins are bound to differ.
    c. Although scientists can move cells into a recipient body, they cannot do this with an entire organ.
    d. The recipient's disease may have progressed to such an extent that the transplant cannot succeed even when matched properly.

11. You and your friend each have a little boy. The kids are always playing in dirt, climbing trees, falling off bikes, and generally getting into everything. When your little Johnny comes home with a runny nose or a small cut, your philosophy is to give him a decongestant or to clean the wound well, and then let it take its own course. Your friend's philosophy is the exact opposite of yours. When her little Tommy comes home with similar problems, she takes him immediately to the doctor for antibiotics. Who is right and who is wrong? Why?
    a. Your friend is right to take her son for antibiotics. One can never know how pathogenic an infectious microorganism can be.
    b. You are. It is probably better to allow his immune system to take care of these minor infections.
    c. Neither of you is doing the best thing. Both of you should vaccinate the boys with every known immunization, thereby protecting them from everything.

12. Which of the following is designed to confer immunity to a particular disease?
    a. antibiotics
    b. neuraminidase inhibitors
    c. vaccines
    d. all of the above

13. If allergic reactions are caused by antibodies binding to allergens, why are the responses localized (limited to a given tissue, such as a runny nose or an upset stomach)?
    a. Antibodies that respond to allergens are attached to mast cells, which are located in tissues, not circulating through the bloodstream.

181

b. Allergens interact directly with tissues and cause inflammatory response wherever they fall.

c. Allergic responses are not tissue specific.

d. None of the answers is correct.

14. Which of the following is a valid treatment for autoimmune diseases?

a. replacement of the lost or damaged tissue

b. antibiotics

c. immune suppression therapy

d. (a) and (b)

e. (a) and (c)

15. Cancer is so hard to cure because cancer cells

a. started as normal body cells that then lost control of cell division, so they are very hard to distinguish from healthy body cells.

b. are more resistant to toxins than normal cells.

c. are genetically different from all of the other cells of the body.

d. None of the answers is correct.

ANSWER KEY

Quiz 1

1. a		**9.** d	
2. b		**10.** a	
3. b		**11.** c	
4. c		**12.** f	
5. c		**13.** b	
6. b		**14.** b	
7. a		**15.** d	
8. b			

Quiz 2

1. b		**9.** b	
2. d		**10.** b	
3. c		**11.** b	
4. a		**12.** c	
5. d		**13.** a	
6. b		**14.** e	
7. c		**15.** a	
8. e			

Chapter 37: CHEMICAL CONTROL OF THE ANIMAL BODY: THE ENDOCRINE SYSTEM

OUTLINE

Section 37.1 How Do Animal Cells Communicate?

- Cells communicate when messenger molecules that bind to target cell **receptors** are released, triggering a change within the target cell.
- Animal cells can communicate directly through gap junctions linking cells, diffusion of chemicals (**local hormones** and **neurotransmitters**) to nearby cells, and by the transport of chemicals (**hormones**) in the bloodstream (**Table 37–1**).

Section 37.2 What Are the Characteristics of Animal Hormones?

- Most cells release **local hormones** that diffuse to nearby target cells (**paracrine communication**).
- **Endocrine hormones** are produced by specialized cells that release them into the bloodstream (**endocrine communication**). There are three classes of hormones: (1) **peptide hormones**, (2) **amino acid derived hormones**, and (3) **steroid hormones** (**Table 37–2**). The **endocrine system** is composed of glands responsible for the production and secretion of hormones into the bloodstream (**Figure 37-1**).
- Hormones bind to specific receptors on **target cells** (**Figure 37-2**). **Peptide** and **amino acid derived hormones** bind to membrane receptors that result in the production of **second messenger** molecules (e.g., cAMP) within the cell. Second messenger molecules initiate an amplified reaction within a cell, resulting in the synthesis or secretion of a desired product (**Figure 37-3**). **Steroid hormones** diffuse through the cell membrane and bind to receptors in the cytosol or nucleus. In the nucleus, the hormone-receptor complex then attaches to a specific gene on DNA, resulting in the production of mRNA that is used to synthesize a protein (**Figure 37-4**).
- Most hormones are regulated by negative feedback, although a few (e.g., oxytocin) are regulated by positive feedback.

Section 37.3 What Are the Structures and Hormones of the Mammalian Endocrine System?

- Mammals have two types of glands: those that release their secretions outside the body (**exocrine glands**) and those that release their secretions inside the body, and these secretions are carried in the bloodstream (**endocrine glands**).
- The **hypothalamus** controls secretions by the **pituitary gland**. The anterior pituitary gland is controlled by hypothalamic **releasing** or **inhibiting hormones**, while the posterior pituitary gland releases secretions from hypothalamic **neurosecretory cells** (**Figure 37-6**).
- The anterior pituitary gland releases seven types of hormones: (1) **follicle-stimulating hormone (FSH)**, (2) **luteinizing hormone (LH)**, (3) **thyroid-stimulating hormone (TSH)**, (4) **adrenocortictrophic hormone (ACTH)**, (5) **growth hormone (GH)**, (6) **prolactin (PRL)**, and (7) **melanocyte-stimulating hormone (MSH)**. The function and chemical type of these hormones are summarized in **Table 37–3**.
- The **posterior pituitary gland** releases hormones produced by hypothalamic neurosecretory cells. **Antidiuretic hormone (ADH)** prevents dehydration by increasing the water permeability of nephron-collecting ducts. **Oxytocin** causes muscular contractions that are involved with childbirth and milk "letdown" (**Figure 37-8**).
- The **thyroid gland** releases the hormones **thyroxine** and **calcitonin** (**Figure 37-9**). **Thyroxine** release elevates metabolic rate (**Figure 37-10**), and **calcitonin** plays a role in calcium uptake from the bloodstream.
- The **parathyroid glands** (**Figure 37-9**) release **parathyroid hormone (PTH)** that elevates blood calcium levels by stimulating calcium release from bone (**Figure 37-11**).
- The **pancreas** controls blood glucose levels through the release of **insulin** and **glucagon**. When blood glucose is high, insulin is released, which stimulates glucose uptake by cells. When blood glucose is low, glucagons are released, which stimulates the breakdown of fat and glucose storage molecules in the liver, resulting in increased blood glucose (**Figure 37-12**).

- The sex organs secrete steroid hormones (sex hormones) that play a key role in puberty and gamete production. **Testes** in males produce **androgens** (e.g., **testosterone**), while **ovaries** in females produce **estrogen** and **progesterone**.
- The **adrenal glands** are composed of two layers: the **adrenal medulla** and the **adrenal cortex (Figure 37-13)**. The **adrenal medulla** releases the hormones **epinephrine** and **norepinepherine** in response to stress and cause changes that prepare the body for emergency action. The **adrenal cortex** releases **glucocorticoids** that help the body cope with short-term stress and that stimulate glucose production. The adrenal cortex also releases **aldosterone**, which regulates sodium content in the blood.
- The **pineal gland** produces **melatonin**, which helps regulate daily rhythms in animals. The **thymus** produces **thymosin**, which stimulates the development of T cells. The kidneys produce **erythropoietin**, which regulates red blood cell production, and **renin**, which helps the body respond to low blood pressure. The heart produces **atrial natriuretic peptide (ANP)**, which reduces blood volume. The stomach and small intestine produce **gastrin**, **secretin**, and **cholecystokinin**, which all help regulate digestion. Fat produces **leptin**, which helps control food consumption.

LEARNING OBJECTIVES

Section 37.1 How Do Animal Cells Communicate?

- Discuss the role of receptor proteins in cell communication.

Section 37.2 What Are the Characteristics of Animal Hormones?

- Describe the functions of local hormones.
- Describe the structure of the three types of endocrine hormones.
- Explain how the second messenger system functions.
- Describe the functions of steroid hormones.
- Discuss the role of negative feedback in hormonal control of the body.

Section 37.3 What Are the Structures and Hormones of the Mammalian Endocrine System?

- Discuss the structural differences between endocrine and exocrine glands.
- Discuss the structural and functional control mechanisms between the hypothalamus and pituitary gland.
- Describe the functions of the anterior and posterior pituitary hormones.
- Explain the functions of the thyroid hormones thyroxine and calcitonin.
- Describe the effects of parathyriod hormone secretion.
- Explain how insulin and glucagon function to maintain blood glucose levels.
- Discuss the roles FSH and LH play in puberty.
- Describe the physiological effects caused by epinephrine and norepinephrine secretion.
- Describe the functions of the adrenal hormones cortisol and aldosterone.
- Discuss the functional difference of melatonin secretion in humans and lower animals.
- Discuss how renin, angiotensin, and atrial natriuretic peptide control blood volume.
- Discuss the effect of thyroxine secretion in different animals.

QUIZ 1

1. You have just discovered a new hormone. When you break the hormone apart to study its structure, you find that it is composed of many amino acids. Which of the following classes of hormones would your new hormone fit into?
   a. steroid
   b. amino acid
   c. peptide
   d. prostaglandin

2. Hormone specificity is determined by receptors on _____.
   a. target cells
   b. endocrine glands
   c. exocrine glands
   d. the hypothalamus

3. Steroid hormones most often exert their action by
   a. entering the nucleus of a cell where they have an effect on the expression of a gene.
   b. finding and binding to the appropriate cell receptor in the cell membrane, which initiates the production of a second messenger.

4. Which of the following represents the "first messenger" when the mechanism of action for a hormone involves the production of a "second messenger"?
   a. mRNA
   b. cAMP
   c. an enzyme activated by an increased level of cAMP in the cytoplasm
   d. the hormone

5. _____ from the hypothalamus stimulate the actions of the anterior pituitary.
   a. Nerve impulses
   b. Exocrine hormones
   c. ADH hormones
   d. Releasing hormones

6. Undersecretion of thyroid hormones can produce _____, characterized by retarded mental and physical development.
   a. cretinism
   b. precocious development
   c. goiter
   d. Graves' disease

7. The hypothalamus responds to decreasing levels of thyroxine by _____.
   a. stimulating the thyroid gland with a nerve impulse
   b. releasing more thyroxine
   c. secreting hormones that stimulate the release of TSH from the anterior pituitary
   d. increasing body temperature

8. The hormone that stimulates the development of specialized white blood cells is _____.
   a. parathormone
   b. estrogen
   c. thymosin
   d. aldosterone

9. Which hormone secreted in a daily rhythm is thought to influence sleep/wake cycles?
   a. insulin
   b. melatonin
   c. thymosin
   d. glucagon

10. Which of the following is characteristic of exocrine glands?
    a. The secretions are released into the bloodstream.
    b. The secretions are not released through a small tube.
    c. They produce hormones.
    d. None of the above is correct.

11. Androgen insensitivity is a rare condition where a genetic (XY) male develops as a female except that the individual has testes that remain in the body cavity and no uterus or ovaries. Blood samples show that the levels of testosterone and other androgens are normal for a male body. Where is the defect in hormonal signaling likely to be in this condition?
    a. LH levels from the anterior pituitary must be below average.
    b. FSH levels from the anterior pituitary must be below average.
    c. Androgen receptors must not be functioning properly and are unable to respond to the circulating androgens.
    d. The cAMP intracellular second messenger pathway must not be functioning properly.

12. Which of the following statements about the role of the hypothalamus is TRUE?
    a. The hypothalamus controls the release and synthesis of hormones by the anterior pituitary through a combination of several hormones, in part because of a unique linking of capillary beds—one in the hypothalamus and one in the anterior pituitary.
    b. The hypothalamic hormones affect the anterior pituitary by stimulating or inhibiting other hormones synthesized by the anterior pituitary.
    c. Modified neurons (neurosecretory cells) with their cell bodies in the hypothalamus, and their endings in the anterior pituitary, release hormones that enter a capillary bed and carry blood to the rest of the body.
    d. (a) and (b) are correct.
    e. All of the above are correct.
    f. None of the above is correct.

13. The concentration of thyroxine in the bloodstream is regulated by
    a. negative feedback mechanisms.
    b. the presence of thyroid-stimulating hormone (TSH).
    c. the presence of thyroid-releasing hormone (TRH).
    d. the parathyroid gland.

14. Which of the following statements about the adrenal gland is true?
    a. The adrenal glands are similar to the pancreas and pituitary in being composed of two different tissues, and therefore the two parts of the adrenal gland have different functions.
    b. The adrenal cortex secretes adrenaline.
    c. The hormones produced by the adrenal cortex are important in the "fight or flight" response.
    d. The adrenal medulla produces three very different functional classes of hormones.

15. If you are resting and your heart rate slows down, which of the following types of communication would be most affected?
    a. paracrine
    b. endocrine
    c. exocrine
    d. autocrine

QUIZ 2

1. Which of the following hormones is based on the structure of cholesterol?
   a. testosterone
   b. glucocorticoid
   c. oxytocin
   d. (a) and (b)
   e. (a) and (c)
   f. all of the above

2. Ibuprofen is often prescribed for relief of menstrual cramps because it
   a. blocks electrical signals from pain nerves to the brain.
   b. acts directly on the muscle cells in the uterus to cause relaxation.
   c. blocks production of prostaglandins.
   d. None of the above is a correct description of how ibuprofen stops menstrual cramps.

3. Why are positive feedback loops much rarer than negative feedback loops?
   a. Positive feedback loops amplify the initial response that triggered them, so without a self-limiting end result, they would be "explosive."
   b. Negative feedback loops control a single variable, while positive feedback loops can control more than one variable. Thus, negative feedback loops are simpler and use less energy.
   c. Negative feedback loops act to negate the response that triggers them, which helps to keep important variables such as body temperature or blood sugar levels within acceptable ranges.
   d. (a) and (b)
   e. (a) and (c)

4. How do peptide hormones exert their specific actions since they circulate throughout the entire body?
   a. The cell secreting the hormone and the cell receiving the hormone are very close together.
   b. The cell receiving the hormone has transmembrane protein receptors that bind to the extracellular hormone.
   c. All cells in the body possess receptors for all circulating hormones, but not all are functional.
   d. None of the above is correct.

5. Insulin is a small (51 amino acid) protein that is produced by the pancreas, which signals the body's cells to take up glucose from the blood. Where is the insulin receptor located in the cell?
   a. in the cell membrane
   b. inside the cell (intracellular)
   c. in the nucleus
   d. none of the above

6. Why does ordinary table salt prevent the development of simple goiters?
   a. Salt is required to stimulate the feedback mechanism that controls the amount of thyroxine produced.
   b. Salts and other minerals inhibit the feedback mechanism, thereby creating the conditions that allow goiters to develop.
   c. Table salt is iodized.
   d. None of the above is correct.

7. Identify the series of events that are initiated when thyroxine levels are low.
   a. The anterior pituitary causes the hypothalamus to secrete TSH, which inhibits the continued production of thyroxine.
   b. The hypothalamus secretes releasing hormones that cause the anterior pituitary to secrete TSH, which stimulates the thyroid to secrete thyroxine.
   c. The thyroid gland produces increased thyroxine when it stimulates the posterior pituitary to secrete its releasing hormones.
   d. The posterior pituitary stimulates the anterior pituitary, which responds by stimulating the thyroid to secrete TSH.

8. Which of the following statements is a correct description of why a particular gland is very large when we are young, but much smaller when we reach adulthood?
   a. The thyroid is most active when we are young to prevent the development of conditions such as cretinism.
   b. The thymus gland is large during youth and decreases with age as our immune system matures.
   c. The pituitary decreases in size after we reach adult stature and growth hormone is no longer needed to stimulate growth.
   d. The pancreas is most active when we are young and decreases in size as less insulin is required when we mature.

9. Gigantism occurs when there is an oversecretion of the hormone that causes exceptionally rapid growth. In some cases, tumor growth can cause oversecretion of hormones. Where would the tumor have to be to give rise to gigantism?
   a. thyroid gland
   b. anterior pituitary
   c. pancreas
   d. adrenal medulla

10. Insulin and glucagon from the pancreas work in concert to keep blood sugar levels under tight control. In addition, hormones such as thyroxine and the glucocorticoids from other glands also play a role in keeping glucose levels in the appropriate range. Why are so many different hormones used to regulate blood glucose levels?
   a. Glucose is used exclusively for energy by all the tissues of the body.
   b. Glucose is the only source of energy the brain can use.
   c. Glucose plays a role in keeping ions inside and outside of cells in balance.
   d. Glucose is used to help carry oxygen in the blood.

11. Exocrine glands release their secretion
   a. to the surface of the skin.
   b. into the bloodstream, where they are distributed throughout the body.
   c. outside of the body and into the digestive tract through tubes or openings called ducts.
   d. as steroid hormones.

12. The posterior pituitary is not a true endocrine gland because it
   a. is only part of the nervous system and has nothing to do with hormones.
   b. is only a hormone storage area and receives the hormones that it releases from the hypothalamus.
   c. has only an exocrine function.
   d. is not involved in homeostasis.

13. Secretions from the _____ are controlled by releasing or inhibiting hormones from the hypothalamus.
   a. anterior pituitary
   b. posterior pituitary
   c. adrenal cortex
   d. thyroid

14. Hypothalamic hormones such as antidiuretic hormone (ADH) and oxytocin that act directly on tissues of the body, must be released from
   a. the anterior pituitary.
   b. the posterior pituitary.
   c. both lobes of the pituitary.
   d. neither lobe of the pituitary.

15. How do hormones from the hypothalamus get to the anterior pituitary gland?
   a. They are secreted from the neurosecretory cells in the hypothalamus directly onto the endocrine cells in the anterior pituitary.
   b. They are released near a capillary bed in the hypothalamus and travel a short distance to a second capillary bed in the pituitary, where they diffuse out around the cells of the anterior pituitary.
   c. They are released near a capillary bed in the hypothalamus and travel around the body before arriving at a second capillary bed in the pituitary, where they diffuse out around the cells of the anterior pituitary.

ANSWER KEY

Quiz 1

1. c
2. a
3. a
4. d
5. d
6. a
7. c
8. c

9. b
10. d
11. c
12. d
13. a
14. a
15. b

Quiz 2

1. d
2. c
3. e
4. b
5. a
6. c
7. b
8. b

9. b
10. b
11. c
12. b
13. a
14. b
15. b

CHAPTER 38: THE NERVOUS SYSTEM AND THE SENSES

OUTLINE

Section 38.1 What Are the Structures and Functions of Neurons?

- Neuron **dendrites** receive electrical signals from other neurons. The neuron **cell body** integrates these signals and generates an electrical signal (**action potential**) that is transmitted down an **axon**. The axon electrical stimulus communicates with other cells at **synapses** that form between **synaptic terminals** and the dendrites or cell body of another cell (**Figure 38-1**).

Section 38.2 How Is Neural Activity Produced and Transmitted?

- If a resting neuron plasma membrane is stimulated, its **membrane potential** becomes less negative, and if **threshold** is reached, an **action potential** will form (**Figure 38-2**). The action potential is conducted down the axon plasma membrane, which occurs more rapidly in **myelinated** axons (**Figure 38-3**).
- When an action potential reaches the **presynaptic membrane** of a synaptic terminal, **neurotransmitters** are released into the **synaptic cleft** of a **synapse**, which then bind to receptors on the **postsynaptic membrane** (**Figure 38-4**). Neurotransmitter binding causes the formation of **inhibitory postsynaptic potentials** (**IPSP**) or **excitatory postsynaptic potentials** (**EPSP**) on the postsynaptic membrane, which undergo **summation** that influences the formation of a new action potential (**Figure 38-4**).

Section 38.3 How Are Nervous Systems Organized?

- All nervous systems must determine the type of stimulus, signal the **intensity** of a stimulus (**Figure 38-5**), integrate information from many sources, and initiate and direct appropriate responses.
- Most behaviors are controlled by neuron-to-muscle pathways composed of **sensory neurons**, **interneurons**, **motor neurons**, and **effectors**. The simplest type of behavior is the **reflex**, which is an involuntary movement of a body part in response to a stimulus (**Figure 38-10**).
- Nervous systems are composed of either a diffuse network of neurons (e.g., a **nerve net**; **Figure 38-6a**) or are centralized (**Figure 38-6b, c**).

Section 38.4 What Is the Structure of the Human Nervous System?

- The vertebrate nervous system consists of the **central nervous system** (**CNS**) and the **peripheral nervous system** (**PNS**; **Figure 38-7**).
- The CNS includes the **brain** and **spinal cord**, while the PNS consists of neurons that lie outside the CNS and the axons that connect these neurons with the CNS.
- The PNS includes **peripheral nerves** that form the **somatic nervous system** (controls voluntary movements) and the **autonomic nervous system** (controls involuntary responses).
- The autonomic nervous system consists of the **sympathetic** and **parasympathetic**, both of which innervate most organs but cause opposite effects (**Figure 38-8**). The sympathetic division helps prepare the body for stressful or energetic activity, while the parasympathetic division directs maintenance activities during periods of rest.
- The central nervous system is protected by the skull and vertebral column, meninges, and the blood-brain barrier.
- The **spinal cord** has nerve axons that extend from the dorsal and ventral portions of the spinal cord (containing sensory and motor neurons respectively; **Figure 38-9**). The center of the spinal cord is made of **gray matter** (composed of motor neuron cell bodies and interneurons) and **white matter** (composed of myelinated axons of neurons that extend up or down the spinal cord).
- The vertebrate **brain** is divided into the **hindbrain**, **midbrain**, and **forebrain** (**Figure 38-11**).
- The **hindbrain** includes the **medulla** and **pons**, which control many involuntary functions, and the **cerebellum**, which coordinates complex motor activities (**Figure 38-12**).

- The human **midbrain** contains the majority of the reticular formation, a filter and relay for sensory stimuli (**Figure 38-12**).
- The **forebrain** includes the **thalamus**, **limbic system**, and the **cerebral cortex** (**Figure 38-12**). The **thalamus** is a sensory relay station that directs information to and from the conscious centers of the forebrain. The **limbic system** (**Figure 38-13**) is a group of structures that work together to produce the most basic emotions, drives, and behaviors. The **cerebral cortex** is the outer layer of the forebrain and is the center for information processing, memory, and initiation of voluntary actions. It is composed of four pairs of lobes: **frontal**, **parietal**, **occipital**, and **temporal** (**Figure 38-14**).

Section 38.5 How Does the Brain Produce the Mind?

- The brain has two **cerebral hemispheres** that are specialized (**Figure 38-15**). The left hemisphere dominates logic, speech, reading, writing, and language comprehension. The right hemisphere specializes in spatial perception, music, emotions, and facial recognition.
- Memory takes two forms: **working memory** and **long-term memory**. **Working memory** is both electrical and biochemical in nature and lasts for only several seconds. Long-term memory involves structural changes in the effectiveness or number of synapses and allows for near-permenant memory retention. The **hippocampus** is important in converting working memory to long-term memory, while the **temporal lobes** are important in the memory recognition of objects and faces and in understanding language.

Section 38.6 How Do Sensory Receptors Work?

- When a **receptor** is stimulated (**Table 38-2**), it generates a **receptor potential** proportional in strength to that of the stimulus. If the receptor potential is strong enough, an action potential will form in a sensory neuron.

Section 38.7 How Are Mechanical Stimuli Detected?

- Mechanical stimuli are detected by mechanoreceptors, which respond to touch, vibration, pressure, stretch, or sound (**Figure 38-17**). Most mechanoreceptors generate receptor potentials when their plasma membrane is deformed or stretched.

Section 38.8 How Is Sound Sensed?

- In the vertebrate **ear**, sound waves vibrate the **tympanic membrane**, which transmits these vibrations to the **middle ear bones** and then to the **oval window** of the **fluid-filled cochlea** that is part of the inner ear (**Figure 38-18a**).
- Within the **cochlea**, vibrations cause the **basilar membrane** to move, bending hairs on **hair cell** mechanoreceptors against the **tectorial membrane** (**Figure 38-18b, c**). The bending of the hairs on hair cells causes action potential formation in auditory nerve axons that lead to the brain.
- **Loudness** is perceived when large vibrations bend hair cell hairs to a greater degree, causing increased neurotransmitter release, resulting in a higher frequency of action potential formation. **Pitch** is perceived based on which hair cells along the cochlear tubes are activated. High-pitch sounds activate hair cells near the oval window, while low-pitch sounds activate hair cells at the tip of the cochlea.

Section 38.9 How Is Light Sensed?

- Arthropods use **compound eyes**, consisting of many light-sensitive **ommatidia** that produce a mosaic image (**Figure 38-20**).
- In the vertebrate **eye**, light enters the **cornea** and passes through the **pupil** to the **lens**, which focuses an image on the **fovea** of the **retina** (**Figure 38-21a**). **Rod and cone photoreceptors** (which are sensitive to light intensity and color wavelengths of light respectively) produce receptor potentials when exposed to light. These receptor potentials are processed by retinal neuron layers (**Figure 38-21b**) that cause in action potential formation in the **optic nerve** that leads to the brain.

Section 38.10 How Are Chemicals Sensed?

- Terrestrial vertebrates detect chemicals in the external environment by **smell** (olfaction) or **taste**. Airborne chemicals are detected by olfactory receptors located in the tissues lining the nasal cavity (**Figure 38-26**). Chemicals that contact the **tongue** are detected by taste receptors located in clusters called **taste buds** (**Figure 38-27**). In both

189

smell and taste, chemical receptors are sensitive to specific types of molecules, allowing for discrimination among odors and tastes.

- **Pain** is a specialized type of chemical sense that involves the detection of chemicals released from damaged tissues near **pain receptors** (**Figure 38-28**).

LEARNING OBJECTIVES

Section 38.1 What Are the Structures and Functions of Neurons?

- Describe the structure of a neuron.
- Describe the four functions of neurons.

Section 38.2 How Is Neural Activity Produced and Transmitted?

- Discuss the difference between the resting potential and threshold.
- Describe the effect myelin has on nerve conduction speed.
- Describe how an action potential is transmitted to another neuron.
- Describe the difference between IPSPs and EPSPs.

Section 38.3 How Are Nervous Systems Organized?

- Discuss how the type of stimulus is determined by the nervous system.
- Discuss the two ways intensity of stimulus information is encoded by the nervous system.
- Describe the path of a reflex.

Section 38.4 What Is the Structure of the Human Nervous System?

- Describe the main divisions of the human nervous system.
- Describe the functions of the somatic nervous system.
- Discuss the difference in function of the sympathetic and parasympathetic nervous systems.
- Describe the structures and main functions of the central nervous system.
- Discuss the specific functions of the medulla, pons, and cerebellum.
- Describe the functions of the thalamus and the limbic system.
- Describe the functions localized to the different lobes of the cerebral cortex.

Section 38.5 How Does the Brain Produce the Mind?

- Discuss the general abilities controlled by the left versus the right hemispheres of the brain.
- Describe the difference between working memory and long-term memory.
- Discuss how long-term memories are made.

Section 38.6 How Do Sensory Receptors Work?

- Describe the different types of vertebrate receptors and the stimuli they respond to.
- Discuss the electrical basis of sensory receptors.

Section 38.7 How Are Mechanical Stimuli Detected?

- Discuss the different stimuli that activate mechanoreceptors.

Section 38.8 How Is Sound Sensed?

- Describe the structures of the outer, middle, and inner ear.
- Describe the pathway of sound waves resulting in hearing.
- Discuss the difference between the perception of volume and pitch.

Section 38.9 How Is Light Sensed?

- Discuss the structure and visual acuity of organisms with compound eyes.
- Describe the structures of the mammalian eye.
- Describe the pathway of light through the mammalian eye from cornea to retina.
- Discuss the difference between the blind spot and the fovea.
- Describe the structural and functional differences between rods and cones.

Section 38.10 How Are Chemicals Sensed?

- Describe how humans are able to detect complex odors.
- Describe the structures relating to our sense of taste.
- Discuss the role of smell in perception of taste.
- Describe the chemical nature of pain perception.
- Discuss the specialized perceptions of sonar, electric, and magnetic fields possessed by some animals.

QUIZ 1

1. Which portion of a neuron conducts impulses toward the cell body?
   a. axon
   b. Schwann cells
   c. dendrites
   d. synaptic terminals

2. The role of the axon is to
   a. integrate signals from the dendrites.
   b. release neurotransmitters.
   c. conduct the action potential to the synaptic terminal.
   d. synthesize cellular components.
   e. stimulate a muscle, gland, or another neuron.

3. When a neuron is maintaining a resting membrane potential
   a. the extracellular (outside the cell) level of sodium ions is higher than that inside the cell.
   b. the inside of the cell is positive relative to the outside.
   c. the cells are more "leaky" to sodium ions than they are to potassium ions.
   d. there is an equal distribution of charge across the membrane.

4. During the first part of the action potential, the inside of the cell becomes less negative because of an
   a. influx of sodium ions.
   b. influx of chloride ions.
   c. influx of potassium ions.
   d. efflux of potassium ions.

5. Action potentials do NOT
   a. represent a brief reversal in the resting potential, so that the inside of the cell becomes positive relative to the outside of the cell.
   b. carry information down the axon to the synapse.

   c. generate when the membrane potential of a cell reaches its threshold potential.
   d. carry information across a synapse.

6. The point at which the action potential is triggered is called the _____.
   a. resting potential
   b. threshold
   c. the repolarization point

7. Depending on the type of synapse, the effect of the neurotransmitter can be _____ to the postsynaptic membrane, making the membrane _____.
   a. excitatory, more negative
   b. inhibitory, less negative
   c. excitatory, less negative

8. Which component of the reflex arc responds to stimuli from the environment and carries it to the central nervous system?
   a. sensory neurons
   b. motor neurons
   c. association neurons
   d. effectors

9. The synapse that transforms incoming information about the environment into outgoing commands to direct a behavior is found in the _____.
   a. peripheral nervous system
   b. dorsal root ganglion
   c. brain
   d. spinal cord

10. Which of these divisions of the peripheral nervous system will innervate skeletal muscle?
    a. autonomic nervous system
    b. sympathetic nervous system
    c. somatic nervous system

11. You are looking at a cross section of the human spinal cord. The gray, butterfly-shaped area is made up mostly of _____.
    a. myelinated axons
    b. cell bodies
    c. meninges

12. In which of its parts does the ear convert the mechanical energy called sound into electrical signals that are sent to the brain?
    a. outer ear
    b. middle ear
    c. inner ear
    d. Eustachian tube

13. The fovea is the
    a. blind spot.
    b. clear area in front of the pupil and iris.
    c. tough outer covering of the eyeball.
    d. substance that gives the eyeball its shape.
    e. central focal region of the vertebrate retina.

14. Which of the following shows the path that light entering the eye and striking the choroid takes?
    a. lens, vitreous humor, cornea, aqueous humor, retina
    b. retina, aqueous humor, lens, vitreous humor, cornea
    c. cornea, aqueous humor, retina, vitreous humor, lens
    d. cornea, aqueous humor, lens, vitreous humor, retina
    e. lens, aqueous humor, cornea, vitreous humor, retina

15. Chemoreceptors for taste are activated when
    a. chemicals dissolved in the saliva bind to receptors on the tongue.
    b. chemicals in air dissolve in mucus and bind to receptors in the olfactory epithelium.
    c. cell contents from damaged tissue stimulate receptor neurons.
    d. none of the above.

QUIZ 2

1. Which part of a neuron is responsible for receiving signals from the environment or from other neurons?
   a. cell body
   b. axon
   c. dendrite
   d. synapse

2. _____ are integration centers in neurons.
   a. Dendrites
   b. Axons
   c. Cell bodies
   d. Ion channels
   e. Synapses

3. During the second part of the action potential, the membrane potential returns to its resting negative state because of an
   a. influx of sodium ions.
   b. influx of chloride ions.
   c. influx of potassium ions.
   d. efflux of potassium ions.

4. To prevent a cell from reaching its threshold potential and generating an action potential, you can
   a. block the potassium leak channels.
   b. open some sodium ion channels.
   c. open some chloride ion channels.

5. How does a drug that causes the neuron's membrane to become a little more permeable to all ions affect the resting and threshold potentials?
   a. The resting and threshold potentials will be unchanged.
   b. The resting potential will move closer to the threshold potential.
   c. The resting potential will move further from the threshold potential.

6. If a neurotransmitter were allowed to remain in the synaptic cleft bound to receptors, the size of the synaptic signal would
   a. remain the same.
   b. be smaller.
   c. be larger.

7. You are recording from a neuron with an electrode and determine that the resting membrane potential is around -70 millivolts. If you added three times more sodium to the extracellular solution, the membrane potential will
   a. stay the same.
   b. become more positive.
   c. become more negative.

8. You have just discovered a new terrestrial species. It is tube shaped, like an earthworm, but it moves at a right angle to the long axis of its body by rolling (like a rolling pin rolling across a table). It has chemoreceptors all over its body, but no obvious eyes or other sensory structures. Which type of nervous system does this new species have?
   a. centralized
   b. nerve net
   c. none

9. How can you distinguish between a light touch and a moderate poke to the arm?
   a. The action potentials are different sizes depending upon the strength of the stimulus.
   b. The frequency of action potentials changes with the strength of the stimulus.
   c. Different neurotransmitters are released depending upon the strength of the stimulus.

10. A patient has suffered a small stroke and reports that he can no longer feel the touch of his hand on a surface. He knows he is touching the surface, but only because he sees his hand there. In which region of the brain did this stroke occur?
   a. frontal lobe
   b. occipital lobe
   c. parietal lobe
   d. hippocampus

11. In which part of the brain might a person have suffered a stroke if her basic emotions and behaviors suddenly change?
   a. limbic system
   b. corpus callosum
   c. occipital lobe
   d. cerebellum

12. Sound reception is carried out by hair cells, which are a special type of
   a. chemoreceptor.
   b. photoreceptor.
   c. mechanoreceptor.
   d. magnetoreceptor.
   e. thermoreceptor.

13. What happens when an image (or a portion of an image) falls on the spot where the ganglion cell axons exit the eye?
   a. Transduction of light into an electrical signal by rods and cones occurs.
   b. Transduction of light energy to electrical energy does not occur here.
   c. Only black and white vision exists at this point.

14. During sudden bursts of intense light, the pupils of the mammal eye constrict because the
   a. aqueous humor will break down when exposed to intense light.
   b. iris constricts to reduce the amount of light projected on the retina.
   c. lens changes shape in an effort to focus the light properly.
   d. ommatidia cannot function properly when exposed to high light intensities.

15. Does having only one tongue present the same disadvantages as if you had only one eye or one ear?
   a. Yes, because having two ears allows us to localize sound.
   b. Yes, because having two eyes gives us a bigger visual range and binocular vision, which again gives us the ability to accurately localize objects in our environment.
   c. Yes, because having two tongues would allow us to better locate food sources in our environment.
   d. No, having two tongues would not allow us to better locate food sources in our environment.

ANSWER KEY

Quiz 1

1. c
2. c
3. a
4. a
5. d
6. b
7. c
8. a

9. d
10. c
11. b
12. c
13. e
14. d
15. a

Quiz 2

1. c
2. c
3. d
4. c
5. b
6. c
7. a
8. b

9. b
10. c
11. a
12. c
13. b
14. b
15. d

Chapter 39: Action and Support: The Muscles and Skeleton

OUTLINE

Section 39.1 An Introduction to the Muscular and Skeletal Systems

- Muscles generate movements through contraction. These movements generate a variety of effects, including locomotion, digestive movements, and the pumping of blood.
- The skeleton provides a framework against which muscles exert force to move the body (**Figure 39-1**). Nearly all animals rely on some form of skeleton for body support.

Section 39.2 How Do Muscles Work?

- Vertebrates have three types of muscle: **smooth**, **cardiac**, and **skeletal**. **Table 39-1** summarizes the structural and functional properties of these muscle types.
- **Skeletal muscle** is composed of bundles of **muscle fibers** (muscle cells) that are attached to bone by **tendons** (**Figure 39-2**). Muscle fibers are composed of **myofibrils** surrounded by a **sarcoplasmic reticulum**. Myofibrils are composed of subunits (**sarcomeres**) attached end to end by **Z line** proteins. Sarcomeres are composed of **thick filaments** and **thin filaments** (**Figure 39-3**).
- Skeletal muscle contraction occurs when motor neurons stimulate skeletal muscle at **neuromuscular junctions** (**Figure 39-5**). This produces an action potential in the skeletal muscle fiber that stimulates the release of calcium ions by the sarcoplasmic reticulum. This causes thin filament accessory proteins to expose the myosin binding sites on **actin** molecules, which in turn allows the thick filament **myosin heads** to bind to actin. Using ATP, the myosin heads bend, release, and reattach, which pulls the thin filaments past the thick filaments (**Figure 39-4**). This causes the sarcomeres to shorten, and thus the muscle fiber shortens.
- The strength and degree of a muscle contraction are determined by the number of muscle fibers stimulated as well as the frequency of action potentials stimulating each muscle fiber.
- Skeletal muscles obtain ATP from cellular respiration (aerobic) or glycolysis (anaerobic). Glycolysis produces ATP rapidly but inefficiently, resulting in the production of lactic acid that contributes to "muscle burn."
- Muscle fibers come in two basic types. **Fast-twitch fibers** contract quickly and powerfully but are poorly adapted for endurance (i.e., aerobic) activities because of their poor blood supply and low mitochondrial density. **Slow-twitch fibers** produce slow and weak contractions but have high endurance because of a rich blood supply and high mitochondrial density that optimizes ATP production by aerobic cellular respiration. Although most humans have equal amounts of both muscle fiber types, the exact proportion is variable and is genetically determined.
- Exertion can cause muscle fibers to grow larger and stronger by adding myofibrils, but the number of muscle fibers is not affected.
- **Cardiac muscle** consists of sarcomeres with myofilaments. Contractions are rhythmic and spontaneous but are synchronized by electrical signals produced by the **sinoatrial (SA) node**. Cardiac muscle cells are interconnected by **intercalated disks** that conduct electrical signals produced by the SA node, resulting in coordinated contraction.
- **Smooth muscle** lacks organized sarcomeres, but its cells are connected by gap junctions that transmit electrical signals from cell to cell. This allows for the production of slow, synchronized involuntary contractions in hollow organs and blood vessels.

Section 39.3 What Does the Skeleton Do?

- Skeletons come in three forms: **hydrostatic skeletons** made of fluid (**Figure 39-6**), **exoskeletons** located on the outside an animal (**Figure 39-7**), and **endoskeletons** located inside an animal (**Figure 39-8**).
- Vertebrate bony endoskeletons perform many functions, including body support, muscle attachment, internal organ protection, blood cell production, and calcium and phosphorous storage. The **axial skeleton** forms the central axis of the body and the **appendicular skeleton** forms the appendages and supporting girdles (**Figure 39-1**).

Section 39.4 Which Tissues Comprise the Vertebrate Skeleton?

- **Cartilage** is composed of **chondrocytes** and the noncellular collagen matrix they secrete (**Figure 39-10**). Cartilage provides flexible support, covers the ends of long bones, and forms shock-absorbing pads in the knee joints and between the vertebrae.
- **Ligaments** are tough bands of collagen that connect bones to each other. This allows the bones at joints to move while remaining attached.
- **Bone** is composed of **osteocytes** and collagen matrix that is hardened by calcium phosphate. Most bones are composed of a hard outer shell of **compact bone** and internal **spongy bone** that may contain bone marrow (**Figure 39-10**). Bone remodeling occurs continuously through the actions of **osteoblasts** (which build bone matrix) and **osteoclasts** (which dissolve bone matrix). When new compact bone is made, osteoblasts deposit concentric layers of bone matrix around channels (containing blood vessels) made by osteoclasts, resulting in the formation of **osteons**.

Section 39.5 How Does the Body Move?

- In the vertebrate skeleton, muscles move the bones at flexible **joints**, which are held together by ligaments. Antagonistic muscle pairs attach to either side of a joint by tendons (**Figure 39-11**). The tendon attachment at the moveable bone of a joint is called the **insertion**, while the tendon attachment at the stationary bone is called the **origin**.
- Contraction of one of the antagonistic muscles bends the joint and stretches out the opposing muscle. At **hinge joints**, **flexor muscles** bend the joint while **extensor muscles** straighten it. Other joints, such as **ball-and-socket joints**, allow for a variety of possible joint movements (**Figure 39-12**).

LEARNING OBJECTIVES

Section 39.1 An Introduction to the Muscular and Skeletal Systems

- Describe the basic functions of the musculoskeletal system.

Section 39.2 How Do Muscles Work?

- Describe the basic functions of the three types of muscle.
- Describe the organization of a muscle from sarcomere to tendon.
- Explain the role of calcium in muscle contraction.
- Describe the interaction between actin, myosin, troponin, and tropomyosin in contraction.
- Explain how the interaction of actin and myosin results in muscle contraction.
- Explain how the nervous system regulates the force and duration of a muscle contraction.
- Discuss the cellular structures of slow-twitch and fast-twitch fibers.
- Describe the differences between cardiac muscle and skeletal muscle.
- Describe the locations and functions of smooth muscle.

Section 39.3 What Does the Skeleton Do?

- Describe hydrostatic and exoskeleton structures.
- Describe the functions of the vertebrate endoskeleton.

Section 39.4 Which Tissues Comprise the Vertebrate Skeleton?

- Discuss the locations and functions of cartilage in the human skeleton.
- Describe the structure and function of compact bone versus spongy bone.
- Discuss the three types of bone cells and their functions.
- Discuss how bone remodeling relates to growth and aging of the skeleton.

Section 39.5 How Does the Body Move?

- Discuss how muscles are attached to the skeleton providing for movement.
- Describe the difference in mobility between hinge joints and ball-and-socket joints.

QUIZ 1

1. If the sliding filament model of contraction is dependent upon the availability of ATP molecules, then why would a dead body undergo *rigor mortis*?
   a. Cross-bridges of myosin cannot pull actin toward the midline of the sarcomere without ATP.
   b. A dead body's cells cannot make ATP.
   c. Cross-bridges cannot turn loose of the binding sites on actin, and thus the muscle cells cannot return to a relaxed state.

2. Sarcomeres are composed of
   a. repeating units of thick and thin filaments.
   b. many myofibrils.
   c. repeating muscle fibers.
   d. many motor neurons.

3. Which type of tissue is involved in the attachment of a skeletal muscle to bone?
   a. ligament
   b. tendon
   c. adipose
   d. cartilage

4. Which function of neuromuscular junctions is NOT shared by neuron-to-neuron synapses?
   a. Summation of electrical stimulation in the postsynaptic cell always occurs.
   b. Changes in postsynaptic cells may be either excitatory or inhibitory.
   c. Postsynaptic cells will always respond with an excitatory postsynaptic potential when stimulated.
   d. Postsynaptic cells do not require summation to generate an action potential.
   e. (a) and (b) are correct.
   f. (c) and (d) are correct.

5. Which muscle type does NOT contain striations (striping) in its cells?
   a. skeletal muscle
   b. cardiac muscle
   c. smooth muscle

6. Muscles contract uncontrollably when exposed to a strong electric shock because the electric shock
   a. directly causes myosin cross-bridges to bind with actin.
   b. is carried throughout the muscle fibers and stimulates the sarcoplasmic reticula to release their calcium stores.
   c. speeds up the rate at which ATP binds to myosin cross-bridges.

7. Functionally speaking, smooth muscle could NOT be used to generate skeletal movements because its contractions
   a. could be too slow for an animal to react to changes in its environment.
   b. would not allow an animal to generate voluntary movements.
   c. are generally best at producing wavelike contractions, which are inefficient for forceful skeletal movements.
   d. All of the above are correct.

8. When calcium ions are released from the sarcoplasmic reticulum to initiate a muscle contraction, to which proteins do they bind?
   a. tropomyosin
   b. myosin
   c. troponin
   d. actin

9. Which of the following cell functions would be MOST affected by the absence of microtubules?
   a. diffusion and osmosis across the membrane
   b. cell movement and cell division
   c. transcription
   d. translation

10. Consider a muscle that is in the middle of a contraction. Imagine that you could keep all of the muscle cells of that muscle flooded with calcium ions and at the same time remove all of the ATP from those cells during the peak of that contraction. Next, the muscle would
    a. remain contracted.
    b. relax.
    c. contract harder.

11. Which skeleton type would be most desirable for a small animal that needed to change its body shape to squeeze into/through small cracks and crevices in the ground?
    a. endoskeleton
    b. exoskeleton
    c. hydrostatic

12. Osteoporosis is a disease in which calcium is removed from bone with a resultant loss of bone mass and strength. Why are women eight times more likely than men to suffer from this disease?
    a. Women's bones are less massive than men's bones.
    b. Women consume more calcium than men.
    c. Estrogen is reduced after menopause.
    d. All of the above.

13. Which cell type directly participates in the remodeling of bone during the growth of an individual?
    a. osteocytes
    b. osteoclasts
    c. osteoblasts
    d. (a) and (b)
    e. (b) and (c)

14. How would significantly reduced or absent cartilage pads in knees affect joint function?
    a. The bones of the joint would move more easily because cartilage tissues would not be in the way.
    b. Joint movements would be painful because there is no cartilage keeping the bones from grinding together.
    c. Joint movement would not be affected.

15. A torn biceps muscle would keep you from
    a. flexing your forearm.
    b. moving your arm at the shoulder.
    c. extending your forearm.

QUIZ 2

1. A striated muscle
    a. has a striped appearance.
    b. is an involuntary muscle.
    c. takes part in muscle contractions of the heart.
    d. is a very large muscle.

2. The bending of the _____ cross-bridges when the energy of ATP is available provides the force of contraction at the molecular level.
    a. myosin
    b. actin
    c. accessory proteins

3. The nervous system controls the strength and degree of muscle contraction because of the
    _____.
    a. number of fibers stimulated
    b. frequency of action potentials
    c. size of action potentials
    d. all of the above
    e. (a) and (b)

4. Another name for a muscle cell is
    a. muscle fiber.
    b. myofibril.
    c. thick filament.
    d. thin filament.

5. Which protein makes up the thick filament of the sarcomere?
    a. collagen
    b. troponin
    c. actin
    d. myosin

6. Which of the following is the BEST description of a cross-bridge?
    a. the connection of the head of the myosin molecules of the thick filament to the actin of the thin filament
    b. deep indentations along the plasma membrane of the muscle fiber (the sarcolemma)

   c. thin lines composed of fibrous protein bands that attach to the thin filaments of the sarcomere and represent the boundary of an individual sarcomere
    d. flattened, membrane-enclosed compartments that contain high concentrations of ions that are released during the contraction of the muscle.

7. Which type of ion is released from the sarcoplasmic reticulum during a muscle contraction?
    a. hydrogen
    b. potassium
    c. calcium
    d. chloride

8. Why does a muscle produce a stronger contraction after extensive weight training, even though the number of muscle cells does not change?
    a. The number of thick and thin filaments increases in each muscle cell, allowing a sarcomere to form more cross-bridges and thus a stronger contraction.
    b. A greater number of action potentials is sent to a trained muscle, thus generating stronger contractions.
    c. A trained muscle has more motor units than an untrained one, resulting in the potential for stronger contractions.
    d. All of the above are correct.

9. What happens to the tropomyosin after calcium is released from the sarcoplasmic reticulum?
    a. It forms a cross-bridge with the myosin of the thick filament.
    b. It rotates out of the way of the myosin binding sites found on the actin molecules of the thin filament.
    c. It rotates around the thin filament and blocks the myosin binding sites found on the actin molecules of the thin filament.
    d. ATP binds to the tropomyosin to provide energy for the contraction.

10. When a molecule of ATP binds to the myosin head while it is attached to the actin of the thin filament, the
    a. myosin forms a cross-bridge.
    b. myosin head detaches from the actin.
    c. myosin head changes its configuration and pulls the actin of the thin filament during the power stroke.
    d. tropomyosin rotates out of the way of the myosin binding sites on the actin of the thin filament.

11. Which type of skeleton is found outside the body?
    a. hydrostatic
    b. endoskeleton
    c. exoskeleton
    d. all of the above

12. John has fallen and broken his femur. Which of the following will deposit new bone to mend the fracture?
    a. osteocytes
    b. osteoblasts
    c. osteoclasts
    d. chondrocytes

13. Bone-dissolving cells are called
    a. chondrocytes
    b. osteoblasts
    c. osteoclasts
    d. osteocytes
    e. erythroblasts

14. Osteons contain all of the following EXCEPT
    a. blood vessels.
    b. intervertebral discs.
    c. osteocytes.
    d. calcium phosphate crystals.
    e. concentric layers of bone.

15. What do you call the muscle attachment on the hip bone (pelvis) if the other end of the muscle is attached to the femur near the knee? Assume the knee is elevated when the muscle contracts.
    a. flexor
    b. extensor
    c. insertion
    d. origin

ANSWER KEY

Quiz 1

1. c
2. a
3. b
4. f
5. c
6. b
7. d
8. c

9. b
10. a
11. c
12. d
13. e
14. b
15. a

Quiz 2

1. a
2. a
3. c
4. a
5. d
6. a
7. c
8. a

9. b
10. b
11. c
12. b
13. c
14. b
15. d

CHAPTER 40: ANIMAL REPRODUCTION

OUTLINE

Section 40.1 How Do Animals Reproduce?

- **Asexual reproduction** involves a single individual that produces genetically identical offspring through repeated mitotic divisions of cells from some part of its body. Asexual reproduction can occur through **budding (Figure 40-1)**, **fission (Figure 40-2)**, or **parthenogenesis (Figures 40-3 and 40-4)**.
- **Sexual reproduction** involves the meiotic production of haploid gametes (**sperm** and **eggs**), which can then fuse through **fertilization** and form a diploid cell. This cell will divide mitotically, producing a diploid individual. Most species have separate sexes, but some are **hermaphroditic (Figure 40-5)**.
- **External fertilization** occurs in water outside the bodies of the parents. It may be synchronized by environmental cues, courtship rituals or **pheromones (Figures 40-6, 40-7**, and **40-8)**.
- **Internal fertilization** occurs within the female body, either by **copulation** or the transfer of **spermatophores (Figure 40-9)**.

Section 40.2 How Does the Human Reproductive System Work?

- Humans copulate and fertilize eggs internally. Sperm and egg cells are produced by paired **gonads**.
- The ability to reproduce occurs at **puberty**, when hormones produced by the hypothalamus (GnRH) stimulate the anterior pituitary to produce LH and FSH. LH and FSH stimulate the gonads to produce either estrogen (by the ovaries) or testosterone (by the testes), which results in development of gametes and secondary sexual characteristics.
- The male reproductive tract is composed of the **testes** (which produce sperm and testosterone) and accessory structures that produce **semen** and facilitate sperm transfer to the female reproductive tract (**Figure 40-11**). These structures are summarized in **Table 40-1**.
- Sperm production (**spermatogenesis**) occurs by meiotic cell divisions within the **seminiferous tubules** of the testes (**Figures 40-12** and **40-13**). Spermatogenesis and testosterone production are controlled by hormones produced by the hypothalamus (GnRH) and anterior pituitary (LH and FSH; **Figure 40-15**).
- The female reproductive tract is composed of the **ovaries** (which produce eggs, estrogen, and progesterone) and accessory structures that accept sperm and nourish the developing **embryo (Figure 40-16)**. These structures are summarized in **Table 40-2**.
- Egg production (**oogenesis**), hormone production, and development of the uterine lining occur in a monthly cycle controlled by hormones produced by the hypothalamus (GnRH), anterior pituitary (FSH and LH), and ovaries (estrogen and progesterone). The process and control of oogenesis is illustrated in **Figure 40-17, 40-18**, and **E40-1**.
- During **copulation**, the erect penis of the male ejaculates semen into the vagina of the female (**Figure 40-19**). The sperm swim through the vagina, uterus, and uterine tubes where fertilization usually takes place.
- During fertilization, enzymes released by sperm acrosomes digest the outer **corona radiata** and **zona pellucida** layers of the unfertilized egg (**Figure 40-20**). This allows a single sperm cell to penetrate the egg and fertilize it.

Section 40.3 How Can People Limit Fertility?

- Permanent **contraception** can be achieved by **sterilization**, either by **vasectomy** or **tubal ligation (Figure 40-21)**.
- Most temporary contraception methods prevent ovulation or prevent sperm and egg from meeting. These methods are summarized in **Table 40-3**.
- **Abortion** removes the embryo from the uterus but is not considered a contraceptive device because it terminates, rather than prevents, pregnancy.

LEARNING OBJECTIVES

Section 40.1 How Do Animals Reproduce?

- Discuss the difference between budding and regeneration in asexual reproduction.
- Discuss the evolutionary benefits provided by sexual reproduction.
- Discuss the temporal and spatial requirements for spawning to be successful.

Section 40.2 How Does the Human Reproductive System Work?

- Describe the meiotic pathway of spermatogenesis.
- Describe the cellular structures involved in spermatogenesis.
- Discuss the hormonal control over spermatogenesis.
- Describe the meiotic pathway of oogenesis.
- Discuss the differences between oogenesis and spermatogenesis.
- Describe the hormonal control over ovulation.
- Discuss where fertilization occurs.

Section 40.3 How Can People Limit Fertility?

- Discuss how surgical sterilization prevents conception in males and females.
- Discuss the various birth control methods with respect to failure rates and protection from sexually transmitted diseases.

QUIZ 1

1. Which type of asexual reproduction involves haploid eggs maturing into haploid adults?
   a. budding
   b. regeneration
   c. fission
   d. parthenogenesis

2. Which of the following events must occur to maximize the likelihood of a successful mating that takes place outside the bodies of the parents?
   a. A special spermatophore must be produced by the male.
   b. Sperm and eggs must be released at the same time.
   c. A single individual must release both eggs and sperm.
   d. Sperm and eggs must be released within the same limited area.
   e. (b) and (d) are correct.

3. Enzymes in the _____ of the sperm digest protective layers that surround the egg.
   a. nucleus
   b. acrosome
   c. midpiece
   d. tail

4. Fertilization usually occurs when the egg is in _____.
   a. one of the ovaries
   b. the cervix
   c. one of the uterine tubes
   d. the uterus

5. A birth control measure for men cuts and ties off the _____, a tube that carries sperm from the testes to the penis.
   a. vas deferens
   b. urethra
   c. epididymis
   d. seminiferous tubule

6. Which of the following sequences of structures represents the pathway of an egg released during ovulation?
   a. ovary, oviduct, fimbriae, uterus
   b. ovary, fimbriae, uterus, oviduct
   c. ovary, uterus, fimbriae, oviduct
   d. ovary, fimbriae, oviduct, uterus

7. Which of the following male structures produces the secretion that makes up the largest volume of semen?
   a. bulbourethral glands
   b. seminal vesicles
   c. epididymis
   d. prostate gland

8. The role of the fimbriae in the female reproductive system is to
   a. help sweep the egg released during ovulation into the uterine tube.
   b. produce and release estrogen and progesterone.
   c. regulate the opening and closing of the uterus where it connects to the vagina.
   d. receive the embryo as it implants.

9. Which of the following descriptions represents the process of ovulation?
   a. The lining of the uterus is shed.
   b. The egg buries itself in the wall of the uterus.
   c. A sperm fuses with an egg.
   d. A follicle in the ovary ruptures and releases an egg.

10. The role of FSH in the first few days of the menstrual cycle is to
    a. stimulate several follicles within the ovaries to develop and mature.
    b. stimulate ovulation.
    c. maintain the corpus luteum at the start of the cycle.
    d. promotes the shedding of the lining of the uterus.

11. Which of the following hormones has an action that is most similar to human chorionic gonadotropin?
    a. FSH
    b. estrogen
    c. LH

12. Why is the release of FSH and LH inhibited during the second half of the menstrual cycle?
    a. The corpus luteum releases LH and FSH, and the release by the pituitary would be redundant.
    b. The corpus luteum releases high levels of estrogen and progesterone, which inhibit the pituitary release of FSH and LH through a negative feedback system.
    c. The corpus luteum releases high levels of estrogen and progesterone, which inhibit the pituitary release of FSH and LH through a positive feedback system.
    d. The pituitary disintegrates in the second half of the cycle, and therefore FSH and LH release becomes inhibited.

13. Which of the following events is timed to the massive "surge" in the amount of LH released from the pituitary?
    a. shedding of the endometrium
    b. disintegration of the corpus luteum
    c. ovulation
    d. fertilization

14. Which birth control technique is the most effective?
    a. birth control pills
    b. IUD
    c. Norplant
    d. abstinence

15. The combination birth control pill works by keeping the levels of estrogen and progesterone in the circulation fairly high during much of the menstrual cycle. Which of the following descriptions explains how this would prevent a pregnancy?
    a. High levels of estrogen and progesterone promote the shedding of the uterine lining, which prevents an egg from implanting.
    b. High levels of estrogen and progesterone inhibit the activity of LH and FSH by interfering with their ability to bind to cells in the ovary.
    c. High levels of estrogen and progesterone promote the release of FSH and LH from the pituitary, which prevents the development of new follicles and inhibits ovulation.
    d. High levels of estrogen and progesterone inhibit the release of FSH and LH from the pituitary, which prevents the development of new follicles and inhibits ovulation.

QUIZ 2

1. Which of the following is an advantage of asexual reproduction over sexual reproduction?
   a. There is no need to find a mate.
   b. Competition for mates is unnecessary.
   c. There is no waste of sperm or eggs.
   d. All of the above are correct.

2. Given the obvious advantages of asexual reproduction, why do so many animals have the capacity for sexual reproduction?
   a. Sexual reproduction is easier than asexual reproduction.
   b. Offspring produced through sexual reproduction are larger and thus are better able to survive in the environment.
   c. Offspring produced through sexual reproduction have the advantage of possessing genetic variability, thus enhancing natural selection processes for the species.
   d. None of the responses is correct.

3. The concentration of _____ represents the hormonal difference between a child and a young person who is going through puberty.
   a. gonadotropin-releasing hormone (GnRH)
   b. FSH
   c. testosterone or estrogen
   d. insulin
   e. (a), (b), and (c)

4. If defective sperm were produced during spermatogenesis because of cell division errors in
mitosis, which cells would MOST likely be to
blame?
   a. interstitial cells
   b. spermatogonia
   c. primary spermatocytes
   d. secondary spermatocytes

5. The inhibition of the production of _____ would
be an effective means of preventing sperm production without affecting testosterone production.
   a. LH
   b. FSH
   c. estrogen
   d. GnRH

6. If defective oocytes were produced during oogenesis because of cell division errors in meiosis II,
which structures would MOST likely be to blame?
   a. oogonia
   b. primary oocyte
   c. secondary oocyte
   d. spermatids

7. During oogenesis, what is the functional advantage
of the production of only one "large" viable egg,
as opposed to four "small" viable eggs?
   a. A large egg is easier for sperm to penetrate.
   b. It takes less energy to produce one large viable
      egg.
   c. A large egg has more cytoplasm and nutrients,
      increasing the chance that a fertilized egg will
      complete development.
   d. None of the above responses is correct.

8. The interstitial cells produce _____.
   a. sperm
   b. ova
   c. testosterone
   d. estrogen

9. The _____ secretes LH and FSH.
   a. anterior pituitary
   b. hypothalamus
   c. posterior pituitary
   d. ovary

10. Which hormone stimulates ovulation?
   a. estrogen
   b. LH
   c. chorionic gonadotropin
   d. progesterone

11. Which male reproductive structure maintains the
optimum temperature required for sperm production?
   a. vas deferens
   b. scrotum
   c. epididymis
   d. seminiferous tubules

12. The ovarian structure that secretes both progesterone and estrogen is the _____.
   a. primary oocyte
   b. secondary oocyte
   c. corpus luteum
   d. immature follicle

13. During a 28-day female menstrual cycle, pituitary
hormones, ovarian hormones, and hormones of the
hypothalamus interact to drive the cyclic changes
in the uterine lining. What might be the outcome if
the ovaries of a young woman were removed during a surgical procedure?
   a. There would be increases in GnRH from the
      hypothalamus and FSH and LH from the pitu-
      itary.
   b. GnRH from the hypothalamus would decrease,
      while the FSH and LH from the pituitary would
      increase.
   c. GnRH from the hypothalamus would increase,
      while the FSH and LH from the pituitary would
      decrease.
   d. GnRH from the hypothalamus and FSH and LH
      from the pituitary would decrease.

14. Which of the following contraceptive methods is
considered to be the LEAST effective?
   a. IUDs
   b. hormonal regulation of ovulation
   c. removal of the penis from the vagina prior to
      ejaculation
   d. condoms

15. If a woman wanted an effective form of contraception that did not affect the physiology of her
endocrine system, which of the following choices
should she consider?
   a. birth control pills
   b. an IUD
   c. Depo-Provera
   d. Norplant

ANSWER KEY

Quiz 1

1. d
2. e
3. b
4. c
5. a
6. d
7. b
8. a

9. d
10. a
11. c
12. b
13. c
14. d
15. d

Quiz 2

1. d
2. c
3. e
4. b
5. b
6. c
7. c
8. c

9. a
10. b
11. b
12. c
13. a
14. c
15. b

CHAPTER 41: ANIMAL DEVELOPMENT

OUTLINE

Section 41.1 How Do Indirect and Direct Development Differ?

- **Indirect development** occurs when a small, sexually immature **larva** hatches from an egg and then undergoes **metamorphosis**, becoming a sexually mature adult with a very different body form (**Figure 41-1**).
- **Direct development** occurs when a newborn animal is sexually immature but resembles a miniature version of the sexually mature adult (**Figure 41-2**). These animals typically produce large, yolk-filled eggs or nourish the embryo within the mother's body.
- In birds, reptiles, and mammals, **extraembryonic membranes** encase the embryo in a fluid-filled space and regulate nutrient and waste exchange between it and the environment. The structure and function of these embryonic membranes are listed in **Table 41-1**.

Section 41.2 How Does Animal Development Proceed?

- Embryo formation begins with **cleavage** of the fertilized egg, eventually forming a **morula** and then a hollow **blastula** (**Figure 41-3a**).
- The blastula forms a **blastopore** through which blastula cells migrate inward. These cells form three embryonic tissue layers (**endoderm**, **mesoderm**, and **ectoderm**) through a process called **gastrulation** (**Figure 41-3b**). The resulting embryo is called a **gastrula** (**Figure 41-3c**).
- The embryonic tissue layers are the developmental basis for adult organ structures, a process called **organogenesis** (**Table 41-2**). The embryo will continue to grow until it is born, when it will grow even larger, achieve sexual maturity, and eventually die.

Section 41.3 How Is Development Controlled?

- All cells of an animal's body contain a full set of genetic information. However, cells are able to **differentiate** by stimulating and repressing the transcription of specific genes (**Figure 41-5**).
- The developmental fate of most cells is sealed during gastrulation through a process called **induction**, in which chemical messages produced by other cells direct differentiation (**Figure 41-6**).
- The position of cells within an embryo may change, a process guided by contact between cell surface proteins and chemical pathways laid out by other cells.
- Short sequences of DNA found within larger genes (**homeoboxes**) are important developmental regulators that cause the formation of specific body parts (**Figure 41-7**). This occurs when homeobox genes initiate the production of amino acid sequences that allow **transcription factors** to bind to specific genes, causing them to be expressed.

Section 41.4 How Do Humans Develop?

- A fertilized egg (**zygote**) develops into a blastocyst and implants in the endometrium (**Figure 41-9**). During implantation, the outer blastocyst layer forms the chorion, while the inner cell mass forms the amnion, yolk sac, and **embryonic disc** (**Figure 41-10**). Gastrulation begins the second week after fertilization.
- During the third week of development, **chorionic villi** extend into the endometrium of the uterus.
- During the fourth week of development, the endoderm forms a tube that will eventually become the digestive tract. The umbilical cord also forms between the embryo and the **placenta** of the mother (**Figures 41-11** and **41-14**).
- During the sixth week of development, the embryo displays prominent chordate features (notochord, tail, gill grooves), rudimentary eyes, and a developing brain (**Figure 41-12**).
- By the eighth week of development, most of the major organs have developed. The gonads form and develop, producing testosterone or estrogen that will affect future embryonic development. The developing embryo is now called a **fetus** (**Figure 41-13**).

- Over the next seven months, the fetus continues to grow and develop, with its organs becoming more functional. The nine months of human embryonic development are summarized in **Figure 41-8**.
- Nine months of fetal development culminate in **labor** and **delivery** (**Figure 41-15**), which occurs through the interplay of estrogen, progesterone, steroid hormones produced by the fetus, oxytocin, and uterine stretching.
- During pregnancy, the **mammary glands** within the mother's breasts grow under the influence of estrogen and progesterone (**Figure 41-16**). After birth, infant suckling stimulates the release of prolactin and oxytocin, which triggers milk secretion and release from the mammary glands.
- **Aging** is a process in which random damage to essential biological molecules (such as DNA) accumulates over time. This impairs the body's ability to repair cell and tissue damage, resulting in the loss of organ function and, eventually, death.

LEARNING OBJECTIVES

Section 41.1 How Do Indirect and Direct Development Differ?

- Describe the stages of indirect development.
- Discuss the advantages and disadvantages to parent and offspring of indirect development.
- Discuss the advantages and disadvantages to parent and offspring of direct development.
- Describe the functions of the four membranes of amniotic eggs.

Section 41.2 How Does Animal Development Proceed?

- Describe how a zygote becomes a blastula.
- Describe the three layers formed by gastrulation and their eventual organization of the organism.

Section 41.3 How Is Development Controlled?

- Discuss how induction leads to cellular differentiation.
- Describe the importance of homeoboxes on cellular differentiation.

Section 41.4 How Do Humans Develop?

- Describe the process of implantation of a mammalian embryo.
- Discuss the major functions of the placenta for both mother and fetus.
- Describe the hormonal interplay between fetus and mother resulting in the onset of labor.
- Discuss the hormones required for milk production.
- Discuss the cellular basis of aging.

QUIZ 1

1. Which form of development results in animals that, generally, live for a longer period of time?
   a. indirect development
   b. direct development

2. Which of the following animals utilizes indirect development?
   a. duck
   b. iguana
   c. ostrich
   d. frog

3. Which embryonic layer gives rise to muscle tissue?
   a. endoderm
   b. ectoderm
   c. mesoderm

4. In animal development, the process of cleavage can BEST be described as
   a. meiosis.
   b. mitosis.
   c. differentiation.
   d. fertilization.

5. Which of the following stages of development is defined by the three embryonic tissue layers (ectoderm, mesoderm, and endoderm)?
   a. gastrula
   b. zygote
   c. morula
   d. blastula

6. Which of the following stages of development is defined by the process of organogenesis?
   a. blastula
   b. morula
   c. early to late embryo
   d. gastrula

7. In order to prove that most cells in the human body contain sufficient DNA to produce a clone, DNA should be harvested from a(an) _____ cell.
   a. sperm
   b. brain
   c. egg
   d. skin
   e. (b) and (d)

8. Which type of embryonic cell would be MOST effective in helping to repair a spinal cord injury?
   a. bone
   b. liver
   c. skin
   d. stem

9. What happens to a cell during differentiation?
   a. It takes on its specialized function by becoming a specific cell type.
   b. It divides to produce a cell with one-half the number of original chromosomes.
   c. It divides to produce a cell with the same number of chromosomes.
   d. Material moves across the cell's membrane as it goes from a region of high concentration to a region of low concentration.

10. Which of the following provides experimental evidence that aging is controlled by DNA?
    a. The number of mitotic divisions that a cell can undergo is directly proportional to the shortening of the telomeres.
    b. An increased ability to repair DNA (e.g., by combating the deleterious effects of reactive oxygen species) is directly proportional to increased life span for a cell.

c. Some "immortal" cancer cell lines have reproduced for decades while maintaining genetically identical cells, indicating that they have found a way to bypass the regulatory mechanisms in the life span of a cell.
    d. All of the above are correct.

11. After the end of the _____ month of development, the embryo is called a *fetus*.
    b. first
    b. second
    c. fifth

12. In human embryonic development, neural tube eventually becomes part of the
    a. digestive tract
    b. liver
    c. heart muscle
    d. brain and spinal cord

13. Which of the following structures develops into the blastocyst?
    a. zygote or fertilized egg
    b. gastrula
    c. morula
    d. late embryo

14. Which of the following defines the embryonic stage of human development?
    a. first 3 weeks
    b. first 8 weeks
    c. first 22 weeks
    d. last 14 weeks

15. Which of the following stages of development represents the earliest time that the human fetus would be viable if born?
    a. 8 weeks
    b. 12 weeks
    c. 5 months
    d. 7 months

QUIZ 2

1. Which of the following statements about reptilian extraembryonic membranes is NOT true?
   a. The chorion helps regulate the exchange of water between the environment and the embryo.
   b. The amnion encloses the embryo in fluid.
   c. The allantois carries blood between the embryo and the placenta.
   d. The yolk sac contains food for use by the embryo.

2. The greatest difference between indirect development and direct development is that _____.
   a. the body form that emerges from an egg of a species that follows indirect development looks very different from the sexually mature adult form
   b. the body form that emerges from an egg of a species that follows direct development looks very different from the sexually mature adult form

c. species that exhibit the indirect line of development are typified by extraembryonic membranes
   d. none of the above

3. Developmentally speaking, what problems would occur in an embryo if gastrulation did NOT happen?
   a. The embryo could not form a small, solid ball of cells.
   b. Cleavage would not occur.
   c. The embryo could not form a hollow, spherical ball of cells.
   d. Organs would not form.

4. The first stage of cleavage is the _____.
   a. gastrulation
   b. blastula stage
   c. morula

5. Which of these germinal layers will form the skin and the nervous system in development?
   a. mesoderm
   b. endoderm
   c. ectoderm

6. Which of these germinal layers will form the digestive tract?
   a. mesoderm
   b. endoderm
   c. ectoderm

7. If an insect requires a specific temperature to become sexually active and El Niño has kept the temperature below that, then the insect would _____.
   a. develop abnormally because of the relationship between temperature and development
   b. develop earlier to make up for the lost time waiting to develop by others of its kind
   c. delay sexual activity until the proper temperature is reached in its environment

8. Cells constantly receive chemical messages from other cells of the body. These messages can alter developmental life by changing _____.
   a. cell membrane function
   b. carbohydrate and fat metabolism
   c. the transcription of genes and thus of proteins made by the cell

9. How can the fate of embryonic cells be determined by chemical messages received from still other embryonic cells?
   a. Induction
   b. Specific surface proteins recognize chemical trails or pathways, causing the cells that bear them to migrate along these pathways.
   c. Cells may adopt different developmental fates, depending upon their response to a concentration gradient of a regulatory substance.
   d. All of the above are correct.

10. If induction within a developing embryo stopped, _____ would be MOST affected.
    a. cleavage
    b. blastula formation
    c. gastrulation
    d. organogenesis
    e. (c) and (d)

11. Which of the following hormones is released by the developing embryo to ensure the maintenance of the corpus luteum in the ovary?
    a. human chorionic gonadotropin (hCG)
    b. luteinizing hormone (LH)
    c. estrogen
    d. follicle-stimulating hormone (FSH)

12. All of the following are true about the fetal umbilical vein EXCEPT that it _____.
    a. carries nutrient-poor blood
    b. carries blood from the fetus to the placenta
    c. is located in the umbilical cord
    d. carries oxygen-rich blood

13. If a woman were unable to produce milk due to an endocrine gland disorder, she might consider prescribed injections of _____ to alleviate the problem.
    a. oxytocin
    b. prolactin
    c. growth hormone

14. Colostrum is particularly beneficial to a newborn baby because it _____.
    a. is especially rich in fat
    b. provides antibodies to the baby that bolster its immune system
    c. is especially rich in lactose
    d. is high in protein
    e. (b) and (d)

15. Which of the following hormones stimulates the release of milk, a process called *milk letdown*?
    a. estrogen
    b. progesterone
    c. prolactin
    d. oxytocin

ANSWER KEY

Quiz 1

1. b
2. d
3. c
4. b
5. a
6. c
7. e
8. d

9. a
10. d
11. b
12. d
13. c
14. b
15. d

Quiz 2

1. c
2. a
3. d
4. c
5. c
6. b
7. c
8. c
9. d

10. e
11. a
12. b
13. b
14. e
15. d

CHAPTER 42: PLANT ANATOMY AND NUTRIENT TRANSPORT

OUTLINE

Section 42.1 How Are Plant Bodies Organized, and How Do They Grow?

- Earth's landscape is dominated by flowering plants (**angiosperms**; **Figure 42-1**) that can be divided into two groups: **monocots** and **dicots** (**Figure 42-2**).
- Angiosperm **root systems** (**Figure 42-3**) allow a plant to anchor itself in the ground, absorb water and minerals from the soil, store food manufactured during photosynthesis, transport substances to and from the shoot, produce hormones, and interact with soil symbionts.
- Angiosperm **shoot systems** are located above ground and are composed of stems, buds, leaves, flowers, and fruits (**Figure 42-1**).
- Plants are composed of **meristematic cells** that can divide and **differentiated cells** that perform specialized functions.
- Plant growth occurs because of cell division and differentiation of **apical meristem** tissue (**Figures 42-1** and **42-15**) and **lateral meristem** issue (**Figure 42-11**). **Primary growth** occurs in apical meristem tissue of the root and shoot tips, which causes an increase in length and the development of specialized structures. **Secondary growth** occurs in lateral meristem tissue along the length of roots and shoots, which causes an increase in root and shoot diameter.

Section 42.2 What Are the Tissues and Cell Types of Plants?

- Plants are composed of three tissue systems: **dermal**, **ground**, and **vascular** (**Figure 42-3**).
- The **dermal tissue system** covers the plant body. It is composed of an outer cell layer of **epidermal tissue** (**Figure 42-4**), as well as the **periderm** that replaces the epidermal tissue of woody plants as they age (**Figure 42-12**).
- The **ground tissue system** consists of three tissue types (**Figure 42-5**). **Parenchyma** tissue cells are living and thin walled and perform many functions. **Collenchyma** tissue cells are living; have thicker, but flexible, walls; and act to support the plant. **Sclerenchyma** tissue cells are dead; have thick, rigid walls; and support the plant body.
- The **vascular tissue system** is composed of two conducting tissues: **xylem** and **phloem**. **Xylem** conducts water and minerals from the roots to the rest of the plant. It is composed of sclerenchyma fibers, **tracheids**, and **vessel elements** (**Figure 42-6**). **Phloem** conducts water, sugars, amino acids, and hormones throughout the plant body. It is composed of sclerenchyma fibers, **sieve-tube elements**, and **companion cells** (**Figure 42-7**).

Section 42.3 What Are the Structures and Functions of Leaves, Roots, and Stems?

- **Leaves** are the major photosynthetic structures of most plants. They are composed of a broad, flat **blade** that is connected to the stem by a **petiole** (**Figure 42-1**).
- The leaf epidermis consists of a layer of nonphotosynthetic cells, a **cuticle**, and **stomata** (with guard cells) that function to reduce water loss while allowing CO_2 entry. The **mesophyll** (beneath the epidermis) consists of photosynthetic parenchyma cells, air spaces, and vascular bundles (**Figure 42-8**).
- **Stems** elevate and support the plant body (**Figure 42-9**). The stem **epidermis**, like that of leaves, reduces water loss while allowing CO_2 to enter. The stem **cortex** and **pith** function to support the stem, store food, and (in some cases) photosynthesize. Stem **vascular tissues** transport water, dissolved nutrients, and hormones.
- As a stem elongates during primary growth, apical meristem cells are left behind at **nodes** and give rise to **leaf primordia** and **lateral buds** (**Figure 42-9**). Under hormonal influence, lateral buds will develop into branches (**Figure 42-10**).
- Stem secondary growth occurs from **vascular cambium** and **cork cambium** cell divisions (**Figure 42-11**). The **vascular cambium** produces secondary xylem and secondary phloem, increasing stem diameter. The **cork cambium** produces waterproof cork cells that cover the outside of the stem.

- **Roots** anchor the plant, absorb nutrients, and store food. Primary growth causes roots to elongate and branch, forming structures consisting of an outer **epidermis**, a **cortex**, and an inner **vascular cylinder**. The root tip consists of apical meristem cells protected by a **root cap** (**Figure 42-15**).
- Cells of the root **epidermis** absorb water and minerals from the soil, often through the use of **root hairs** (**Figure 42-17**). The root **cortex** is primarily responsible for the storage of sugars in the form of starch, although **endodermis cells** control the movement of water and minerals from the soil into the vascular cylinder (**Figure 42-18**). The root **vascular cylinder** contains conducting tissues that receive water and minerals from the cortex endodermis and then transport them to the rest of the plant.

Section 42.4 How Do Plants Acquire Mineral Nutrients?

- Most minerals are actively transported into root hair cells and then diffuse through the cortex to the **pericycle cells** just inside the vascular cylinder. The pericycle cells actively transport minerals to the extracellular fluid of the vascular cylinder, where they then diffuse into the tracheids and vessel elements of xylem (**Figure E42-1**).
- **Fungal mycorrhizae** help convert inaccessible soil minerals into forms that the root can absorb, while the fungus obtains sugars, amino acids, and vitamins from the plant (**Figure 42-19**).
- Legumes have formed a symbiotic relationship with **nitrogen-fixing bacteria**, that allow them to obtain nitrogen from normally inaccessible atmospheric sources (**Figure 42-20**).

Section 42.5 How Do Plants Move Water Upward from Roots to Leaves?

- The **cohesion-tension theory** explains how water moves from the roots to the leaves (**Figure 42-21**).
- When water evaporates from the leaves through stomata (**Figure 42-22**), water within the xylem enters the leaves to replace the lost water.
- Cohesion of water molecules to each other (and the xylem walls) creates a "chain" of water molecules that runs down the length of the xylem plant roots.
- Transpiration of water from the leaf creates "tension" that pulls the chain of water molecules toward the leaf, as well drawing water into the vascular cylinder of the root from surrounding endodermis cells.

Section 42.6 How Do Plants Transport Sugars?

- The **pressure-flow theory** explains how sugar moves in phloem (**Figure 42-24**).
- Sugar is produced in **source cells** (e.g., leaves) and is actively transported into phloem tubes, which causes the movement of water (by osmosis) into the phloem from adjacent xylem tubes.
- The increase in water volume causes an increase in fluid pressure that drives fluid movement (and dissolved sugars) into regions of lower pressure.
- Cells in **sugar sinks** (e.g., fruit or roots) actively transport sugars out of phloem tubes, with water following by osmosis. Thus, a fluid pressure gradient is established between source cells and sink cells that perpetuates sugar movement.

LEARNING OBJECTIVES

Section 42.1 How Are Plant Bodies Organized, and How Do They Grow?

- Discuss the parts common to angiosperms.
- Explain the difference between monocots and dicots.
- Describe the major functions performed by roots.
- Describe the stem structures and functions.
- Discuss the difference between meristematic cells and differentiated cells in plants.
- Discuss the difference between primary growth, and secondary growth.

Section 42.2 What Are the Tissues and Cell Types of Plants?

- Describe the three types of tissues found in plants.
- Describe the varied functions of epidermal tissues.
- Describe the structure and functions of parenchyma, collenchyma, and sclerenchyma.
- Describe the structure and function of xylem and phloem.

Section 42.3 What Are the Structures and Functions of Leaves, Roots, and Stems?

- Describe the structure and function of the blades and petioles of leaves.
- Describe the structure and functions of stomata, guard cells, and mesophyll.
- Describe the various functions of plant stems.
- Describe the organization of cells in the stem that provide for growth, nutrient storage, and transportation.
- Discuss the formation and growth of nodes and lateral buds.
- Describe secondary growth in dicots.
- Explain how tree rings are formed.
- Discuss the functional difference between sapwood and heartwood.
- Describe the formation and function of cork.
- Describe the difference between taproot and fibrous root systems.
- Describe the formation and function of the root cap.
- Describe the formation of a branch root.
- Describe the formation and function of root hairs.
- Discuss the importance of the casparian strip.
- Describe the pathway of water from soil to pericycle cells.

Section 42.4 How Do Plants Acquire Mineral Nutrients?

- Discuss which minerals plants require and how they obtain them.
- Discuss the role of micorrhizae in nitrogen fixation.
- Discuss the mutually beneficial relationship resulting in nitrogen fixation.

Section 42.5 How Do Plants Move Water Upward from Roots to Leaves?

- Discuss the interaction of cohesion and tension that results in the upward movement of water.
- Discuss the role of stomata and guard cells in water transportation.
- Explain the various ways plants regulate their stomata.

Section 42.6 How Do Plants Transport Sugars?

- Explain the pressure flow theory for sugar transport.
- Discuss leaf adaptations with respect to water availability.

QUIZ 1

1. When you were a 3-foot-tall child, you drove a nail in a tree trunk at about eye level to hang a toy on. Now it is 10 years later and you are 6 feet tall. The tree has grown 1 foot a year. How high will the nail be on the side of the tree?
   a. 3 feet
   b. 6 feet
   c. 9 feet
   d. 12 feet

2. Primary growth, but NOT secondary growth,
   a. is produced by only lateral meristems.
   b. is produced by either lateral or apical meristems.
   c. increases the length of stems and roots.
   d. increases the diameter of stems and roots.
   e. occurs in stems but not in roots.

3. A region of actively dividing cells (e.g., those at the tip of the root or stem) in plants is referred to as _____.
   a. parenchyma
   b. vascular tissue
   c. ground tissue
   d. a meristem

4. A tree trunk increases in size. What is this process called, and what are the mechanisms responsible for such growth?
   a. primary growth, mitotic cell division and cell differentiation
   b. secondary growth, mitotic cell division and cell differentiation
   c. differentiated cells, undifferentiated meristem
   d. apical meristem, mitosis

5. Which main type of ground tissue would you be ingesting if you ate a carrot?
   a. parenchyma
   b. collenchyma
   c. sclerenchyma
   d. equal part of (a), (b) and (c)

6. Which tissue in the above-ground portion of a plant body contains the least densely packed cells? What is the function of the air spaces between the cells?
   a. collenchyma, source of oxygen for photosynthesis
   b. pith, helps with support
   c. xylem, allows pressure for transport
   d. spongy and palisade parenchyma, source of carbon dioxide for photosynthesis and oxygen for cellular respiration

7. The primary photosynthetic cells of dicot leaves are _____ cells.
   a. epidermal
   b. guard
   c. mesophyll parenchyma
   d. collenchyma

8. Why would one NOT expect to find root hairs in the area between the region of cell elongation and the root cap?
   a. There are no dermal cells in this area.
   b. The root cap is pushed down between soil particles by cells undergoing elongation. Root hairs would be torn off. Root hairs are formed later in development in the mature region.
   c. There is no vascular cylinder in these regions.
   d. These regions of the root do not require water.

9. The majority of water and dissolved minerals absorbed by a plant is first taken up by the
   a. root hairs.
   b. cortex.
   c. endodermis.
   d. vascular tissue of the root.

10. Which of the following drives the movement of water from the soil into the central root?
    a. transpiration
    b. active transport
    c. osmosis
    d. all of the above

11. Water leaves the plant through
    a. root hairs.
    b. stomata.
    c. sieve-tube elements.
    d. the epidermis.

12. The cohesion-tension mechanism of water movement occurs when _____.
    a. the buildup of water pressure in the roots forces water up the stem through the xylem
    b. the water is loaded into vessel elements by active transport at the source and unloaded at the sink to ensure a constant movement of water through the xylem

    c. the transpiration of water from the leaf surface pulls water up the xylem as a result of hydrogen bonding between water molecules
    d. the stem contracts and decreases its diameter to force the water molecules through the tracheids and vessels

13. While visiting your aunt in Michigan, one morning you find a beaver gnawing on a birch tree. You scare it off but not before it has removed a strip of bark all the way around the tree. Despite erecting a fence around the tree to prevent additional damage, the tree dies after a few months. The first event that would occur would be the death of the _____.
    a. leaves of the tree because they stopped receiving water from the soil
    b. leaves of the tree because they could no longer produce sugars by photosynthesis
    c. roots of the tree because they would no longer be able to absorb water from the soil
    d. roots of the tree because they would no longer be able to receive carbohydrates from the leaves

14. Maple syrup is produced from fluid taken from sugar maple trees in the late winter or very early spring. These trees are tapped only during this time of the year because this is
    a. when roots serve as a source, and young buds are the sink.
    b. the time of year roots are a sink.
    c. the only time when sugars are transported.
    d. when cold temperature allow sapwood to transport fluids.

15. Pressure-flow theory explains sugar movement in phloem from _____, where sugars are actively made, toward _____, where sugars are used or converted to starch.
    a. sink, source
    b. sink, sink
    c. source, sink
    d. source, source

QUIZ 2

1. Which of the following does NOT identify a plant as a dicot?
   a. leaves with veins arranged like a net
   b. hand-shaped leaves
   c. six petals
   d. a taproot
   e. a seed with two cotyledons

2. The lateral bud can give rise to
   a. leaves and flowers.
   b. roots and branches.
   c. branches and flowers.
   d. leaves and roots.

3. The portion of the stem where the leaves are attached is called the _____; the portion of the stem that is free of leaves is called the _____.
   a. junction, interjunction
   b. node, internode
   c. bud site, interbud site
   d. pith, interpith

4. In a tree that is several years old, it is usually impossible to find any of the primary tissues that were outside the vascular cambium in the primary plant body. Those tissues
   a. were absorbed by the vascular cambium.
   b. became part of the secondary phloem.
   c. fragmented and fell off the outside of the stem.
   d. are now in the center of the stem.

5. In the phloem tissue, the cells that conduct water, sugars, amino acids, and hormones throughout the plant are called
   a. sieve-tube elements.
   b. companion cells.
   c. tracheids.
   d. vessel elements.

6. Xylem is used to
   a. transport sugars down the plant.
   b. grow new plant tissues.
   c. transport water and minerals up the plant.
   d. regulate the flow of water out of the plant.

7. The primary function of the vascular tissue system in plants is to _____.
   a. transport water, minerals, and sugars throughout the plant
   b. store mineral and food reserves for the plant
   c. cover and protect the outer surfaces of the plant
   d. perform photosynthesis within the plant body

8. One function of the spongy mesophyll layer is
   a. protecting the plant tissues.
   b. transporting nutrients.
   c. absorbing water.
   d. gas exchange.

9. A structure between a plant root and a fungus that helps the plant attain nutrients from the soil is _____.
   a. mycorrhizae
   b. root nodules
   c. root hair
   d. lichen

10. The evaporation and loss of water vapor through the leaves of a plant is called _____.
    a. photosynthesis
    b. respiration
    c. cohesion
    d. transpiration

11. Which of the following is NOT a step by which minerals become absorbed by roots?
    a. diffusion of minerals into root hairs
    b. diffusion through cytoplasm to pericycle cells
    c. active transport into the extracellular space around the xylem
    d. diffusion into the xylem

12. Companion cells appear to help regulate _____ transport.
    a. mineral
    b. sugar
    c. water

13. If a hollow needle were inserted into the phloem of a plant
    a. air would be drawn into the cell.
    b. sugary sap would be pushed out of the cell.
    c. pure water would be pushed out of the cell.
    d. nothing would happen.

14. How do plants overcome the force of gravity and make water flow upward?
    a. transpiration of water from the leaves
    b. hydrogen bonds between water molecules
    c. both of the above
    d. neither of the above

15. How do plants balance the need to photosynthesize, which requires carbon dioxide from the atmosphere, with the water loss that comes from transpiration?
    a. by regulating the opening and closing of the stomata
    b. by regulating the potassium ion concentration within the guard cells
    c. by increasing potassium concentrations when photosynthesis outstrips cellular respiration because of the influence of low carbon dioxide levels under these conditions.
    d. all of the above

ANSWER KEY

Quiz 1

1. a
2. c
3. d
4. b
5. a
6. d
7. c
8. b

9. a
10. d
11. b
12. c
13. d
14. a
15. c

Quiz 2

1. c
2. c
3. b
4. c
5. a
6. c
7. a
8. d

9. a
10. d
11. a
12. b
13. b
14. c
15. d

CHAPTER 43: PLANT REPRODUCTION AND DEVELOPMENT

OUTLINE

Section 43.1 What Are the Basic Features of Plant Life Cycles?

- The sexual life cycle of plants involves an **alternation of generations** (**Figure 43-2**) between a multicellular diploid form (**sporophyte** generation) and a multicellular haploid form (**gametophyte** generation).

Section 43.2 How Is Reproduction in Seed Plants Adapted to Drier Environments?

- A male gametophyte is surrounded by a watertight coat, forming a microscopic pollen grain that can withstand desiccation and be dispersed by the wind or animals (**Figure 43-3**).
- The female gametophyte remains moist within the flower of a plant. Upon fertilization by pollen, the egg is enclosed in a drought-resistant seed that contains a food reserve for the embryonic plant.

Section 43.3 What Is the Function and Structure of the Flower?

- **Flowers** are the reproductive structures of angiosperms, most of which lure animals that directly pollinate them.
- Complete flowers have four parts: **sepals**, **petals**, **stamens**, and **carpels** (**Figure 43-4**). **Sepals** form the outer protective covering of the flower bud and are found at the base of a flower. **Petals** can be brightly colored and attract pollinators to the flower. A **stamen** consists of a **filament** bearing an **anther** that produces pollen. The **carpel** consists of an **ovary** (in which female gametophytes develop) and a **style**, at the tip of which is a sticky **stigma** to which pollen adheres during pollination.
- **Pollen** (**Figure 43-8**) contains the male gametophyte that develops within the anther of the sporophyte plant (**Figure 43-7**). Diploid **microspore mother cells** meiotically form four haploid **microspores** that meiotically produce a haploid male gametophyte. The male gametophyte usually consists of a **tube cell** and a **generative cell**, the latter of which forms two sperm cells.
- The **ovary** contains the female gametophyte (**ovule**) that will form **egg cells** (**Figure 43-10**). Within the ovule, a diploid **megaspore mother cell** meiotically forms four haploid **megaspores**, of which only one survives. The surviving megaspore mitotically produces eight haploid nuclei, which eventually form seven haploid cells. The **central cell** is larger than the other six and has two nuclei. One of the cells near the ovule opening will become the **egg cell**.
- **Pollination** leads to **fertilization** of the egg cell and occurs when a pollen grain lands on a stigma (**Figure 43-11**). When this happens, the pollen tube cell grows through the style to the female gametophyte. The generative cell forms two sperm cells that travel down the style (through the tube cell). One of these sperm cells will fertilize the egg cell, forming a diploid zygote that will become an embryo. The other sperm cell will fertilize the central cell, producing a triploid cell that will form the endosperm of the seed.

Section 43.4 How Do Fruits and Seeds Develop?

- **Fruit** develops from the ovary and is modified to encourage seed dispersal (**Figure 43-12**). Many fruits are sweet to encourage animal consumption, have hooks to so they can be carried and dispersed on animals, or have wings for wind dispersal.
- The **seed** contains an **embryo**, consisting of an **embryonic root** and **embryonic shoot**, the latter of which contains the **cotyledon** (one in monocots, two in dicots; **Figure 43-13**). Cotyledons absorb food from the endosperm and transfer it to the growing embryo.

Section 43.5 How Do Seeds Germinate and Grow?

- Seed **germination** requires warmth and moisture but may also require periods of drying, cold, or seed coat weathering. If these conditions are not met, the seed may remain dormant for an extended period of time.

- During germination (**Figure 43-14**), the root emerges from the seed and absorbs water and nutrients. The embryonic shoot consists of a **hypocotyl** and **epicotyl**, which dicots use to disrupt the soil and form a path for the root apical meristem. Monocots use a **coleoptile** that encloses the shoot, allowing it to push aside the soil as the shoot emerges.
- **Cotyledons** supply the developing plant with food energy as it grows (**Figure 43-15**).

Section 43.6 What Are Some Adaptations for Pollination and Seed Dispersal?

- Plants and their animal pollinators and seed dispersers have coevolved over time, acting as agents of natural selection for one another. Flowers attract animals with scent, food, or appealing colors (**Figures 43-18** and **43-19**). Other flowers deceive pollinators with sexual attractants (**Figure 43-20**). Some plants and pollinators are codependent, with some plants providing nurseries for their pollinators.
- Fruits can disperse seeds by shotgun dispersal, being carried on wind or water (**Figures 43-22** and **43-23**), clinging to animal fur (**Figure 43-24**), and enticing an animal to eat the fruit (thus carrying the seed in its digestive tract; **Figure 43-25**).

LEARNING OBJECTIVES

Section 43.1 What Are the Basic Features of Plant Life Cycles?

- Discuss the benefit of sexual over asexual reproduction.
- Discuss the difference between the sporphyte and the gametophyte generations.

Section 43.2 How Is Reproduction in Seed Plants Adapted to Drier Environments?

- Discuss the adaptations of eggs and sperm to overcome the lack of water for fertilization on dry land.
- Describe the steps of fertilization in conifers.

Section 43.3 What Is the Function and Structure of the Flower?

- Explain the concept of insect pollination.
- Discuss the adaptations of plants to encourage insect pollination.
- Discuss the formation and locations of the sperm and eggs in a flowering plant.
- Explain the mechanics of fertilization in flowering plants.
- Describe the four parts of complete flowers.
- Describe the male and female reproductive structures of flowers.
- Describe the structure and formation of a pollen grain.
- Describe egg formation.
- Discuss the difference between pollination and fertilization.
- Describe the fate of each sperm cell after pollination.

Section 43.4 How Do Fruits and Seeds Develop?

- Discuss the formation and function of a fruit body over the seed.
- Discuss the formation and function of cotyledons.

Section 43.5 How Do Seeds Germinate and Grow?

- Discuss the advantages of seed dormancy.
- Explain the environmental requirements to break seed dormancy.
- Describe the steps in germination.

Section 43.6 What Are Some Adaptations for Pollination and Seed Dispersal?

- Describe the mutually beneficial relationship between flowers and their insect pollinators.
- Discuss plant adaptations specific to their pollinators.
- Discuss the variety of fruit dispersal methods employed by plants.

QUIZ 1

1. In the life cycle of a plant, spores are produced by the _____.
   a. gametes
   b. sporophyte
   c. eggs
   d. gametophyte

2. All plant life cycles consist of two unique stages, one in which the chromosomal number of cells is haploid and the other in which the chromosomal number of cells is diploid. Which haploid plant cell undergoes repeated mitotic divisions to produce a haploid adult organism?
   a. sporophyte
   b. gametophyte
   c. spore
   d. gamete

3. The principal function of the petals of a flower is to _____.
   a. produce the female gametophyte
   b. produce pollen grains
   c. attract pollinators
   d. protect the flower body

4. The embryo sac is contained within the _____, which is located inside the _____.
   a. anther, stamen
   b. stigma, carpel
   c. ovule, ovary
   d. ovary, ovule

5. The male gametes of flowering plants are contained within the _____.
   a. seed
   b. stigma
   c. ovule
   d. pollen grain

6. Which of the following is TRUE of double fertilization?
   a. One sperm fuses with the egg to form a zygote, and a second sperm fuses with the polar nucleus to form the endosperm.
   b. One sperm fuses with the tube cell to form the embryo, and a second sperm fuses with a generative cell to form the endosperm.
   c. The first fertilization event results in the production of a zygote, and a second fertilization event occurs several days or weeks later to produce a second zygote, which degenerates to the endosperm.

   d. Double fertilization is a rare event in flowering plants that results in the production of two identical embryos.

7. Which type of cellular division occurs within the pollen sac of the anther?
   a. binary fission
   b. conjugation
   c. mitosis
   d. meiosis

8. A pollen grain has three different nuclei. The tube cell nucleus is one. Of the other two nuclei, one will fertilize the egg and the other
   a. will fertilize a synergid.
   b. will direct the growth of the pollen tube.
   c. is used as a spare.
   d. will unite with the polar nuclei.

9. Fruit flesh is derived from the _____ and functions to _____.
   a. ovary, supply nutrients to the growing embryo
   b. endosperm, protect the developing embryo
   c. ovule, supply nutrients to the growing embryo
   d. ovary, aid in seed dispersal

10. Many "vegetables" such as squash and tomatoes are actually fruits because they
    a. are too sweet to be true vegetables.
    b. are too brightly colored to be vegetables.
    c. develop from the ovary of the flower.
    d. develop from the ovule in a flower.

11. If a fire is hot enough to kill all of the existing plants in an area, new plant life is usually established in the area within a year. The first plants are commonly grasses and annual dicots. How can the rapid establishment of these plants be explained?
    a. The seeds of these plants were present in the soil all along and were released from their dormancy by the heat of the fire.
    b. The fruits of these plants were transported into this area from outlying, unburned regions.
    c. The seeds of these plants were near the surface of the soil at the time of the fire and were protected by their seed coat and dormancy. Their dormancy was broken by their new exposure to the light when the covering plants were burned away.
    d. All of the above are correct.

217

12. What advantage does dormancy (characterized by decreased metabolic activity and resistance to adverse environmental conditions) provide many recently matured seeds?
    a. It provides an enforced delay before germination can begin upon maturation of the seed.
    b. It can provide sufficient time for the dispersal of seeds from the parent plant by weather and/or animals.
    c. It can provide an enforced delay until environmental conditions that are favorable for the growth of the seed exist.
    d. All of the above are correct.
    e. None of the above is correct.

13. What are some of the mechanisms that plants have developed to protect growth of structures during germination?
    a. covering the delicate tips of structures growing through soil with protective caps
    b. extending the period of dormancy
    c. allowing the delicate tip of a growing shoot "trail in the wake" of another structure that is "armored" against the abrasive effects of soil particles
    d. (a) and (c)
    e. (b) and (c)

14. The first leaflike structures to emerge from the seed are
    a. true leaves.
    b. not true leaves but cotyledons.
    c. not true leaves but part of the coleoptile.

15. A method flowers may use to attract pollinators is the production of
    a. nectar.
    b. a rotten odor.
    c. pollen.
    d. all of the above.

QUIZ 2

1. In plants, the gametophyte produces eggs and sperm by
    a. mitosis.
    b. meiosis.
    c. spore formation.
    d. fertilization.

2. When you are gazing at a lovely magnolia tree, which generation of this flowering plant are you looking at?
    a. sporophyte
    b. gametophyte

3. Which of the following is the outermost structure of the flower?
    a. stamen
    b. stigma
    c. sepals
    d. ovary

4. Which of the following is specialized to receive pollen?
    a. stamen
    b. stigma
    c. sepals
    d. ovary

5. Which of the following is NOT part of the carpel?
    a. anther
    b. style
    c. ovary
    d. stigma

6. Before it is carried to the flower of another plant, the pollen grain is created within the _____, which is part of the _____.
    a. anther, stamen
    b. stigma, carpel
    c. ovule, ovary
    d. ovary, ovule

7. Flowers are made of modified _____.
    a. sepals
    b. leaves
    c. stigmas
    d. fruits

8. What happens to the megaspore mother cell contained inside an ovary?
    a. Nothing. The megaspore is eventually fertilized by a pollen grain.
    b. The megaspore cell develops into a series of layers called *integuments*.
    c. The megaspore cell divides meiotically to produce four haploid megaspores.
    d. None of the above is correct.

9. The integuments mature to become the
    a. fruit flesh.
    b. seed coat.
    c. food for the embryo.

10. A seed contains each of the following EXCEPT a(n) _____.
   a. embryonic plant
   b. food source
   c. seed coat
   d. stoma

11. The first structure to emerge from the seed is (are) the
   a. cotyledons.
   b. root radicle.
   c. hypocotyl.
   d. leaves.

12. The endosperm of the seed functions to
   a. provide nutrients to the growing embryo.
   b. develop into the new plant.
   c. provide protection for the growing embryo.
   d. aid in seed dispersal.

13. The function of the cotyledons in most dicot seeds is to _____.
   a. store food for the embryo and growing seedling
   b. protect the embryo

   c. photosynthesize to provide food for the dormant embryo
   d. help with seed dispersal

14. When a bee gets nectar from a flower, the bee picks up pollen from the _____ and carries it to another flower.
   a. stigma
   b. ovary
   c. sepals
   d. anthers
   e. filament

15. Which of the following is a way that flowering plants disperse their seeds?
   a. tasty fruits that entice animals to eat them
   b. sticky burrs that attach to animal fur
   c. lightweight structures that are carried by wind
   d. all of the above

ANSWER KEY

Quiz 1

1. b
2. c
3. c
4. c
5. d
6. a
7. d
8. d
9. d
10. c
11. d
12. d
13. d
14. b
15. d

Quiz 2

1. a
2. a
3. c
4. b
5. a
6. a
7. b
8. c
9. b
10. d
11. b
12. a
13. a
14. d
15. d

CHAPTER 44: PLANT RESPONSES
TO THE ENVIRONMENT

OUTLINE

Section 44.1 What Are Plant Hormones, and How Do They Act?

- Plant **hormones** are produced by cells in one location and transported to other parts of the body, where they exert specific effects.
- The five major plant hormone categories are **abscisic acid**, **auxins**, **cytokinins**, **ethylene**, and **gibberellins**. The functions of these hormones are listed in **Table 44-1**.

Section 44.2 How Do Hormones Regulate the Plant Life Cycle?

- **Abscisic acid** maintains seed dormancy and must be washed away by moisture before growth occurs. (**Figure 44-3**).
- **Gibberellin** stimulates germination, initiating the synthesis of enzymes that digest food reserves.
- **Auxin** stimulates seedling shoot growth toward light (**phototrophism**) and root growth against gravity (**gravitrophism**). This is accomplished by the stimulation of cell elongation because of the accumulation of auxin along internal cell layers of the shoot/root (**Figure 44-4**). Young plants may sense gravity because of the accumulation of plastids along the interior of cell surfaces (**Figure 44-5**).
- The shape of a mature plant is determined by the interplay between auxin and cytokinin. The ratio of auxin (which inhibits lateral bud formation and is produced by the shoot tip) to cytokinin (which stimulates lateral bud formation and is produced by the root tip) controls lateral bud (and branch) formation (**Figure 44-7**). The phenomenon of the shoot tip dominating the activity of lateral buds is referred to as **apical dominance** (**Figure 44-6**).
- The timing of flowering is typically controlled by the duration of darkness (**Figure 44-8**), which may stimulate the production of **florigens**.
- Plants can detect light and darkness by changes in a leaf pigment called **phytochrome** (**Figure 44-9**). Phytochrome influences plant responses to light, including flowering, seedling elongation (**Figures 44-10** and **44-11**), leaf growth, chlorophyll synthesis, and straightening of the epicotyl or hypocotyl hook in dicots.
- Developing seeds produce auxin and/or gibberellin into the surrounding ovary tissues, which stimulates fruit production.
- As a seed matures, a surge in auxin concentration stimulates ethylene production, which causes fruit ripening (**Figures 44-12** and **44-13**). During ripening, starches are converted into sugars, the fruit softens, and colors brighten.
- Temperate plants prepare themselves for winter by undergoing **senescence**, due primarily to a fall in auxin and cytokinin levels. This process of rapid aging results in the formation of **abscission layers** at fruit and leaf petioles (**Figure 44-14**). Simple sugars are transported to the roots for winter storage and plant buds become dormant, a condition maintained by high levels of abscisic acid.

Section 44.3 Can Plants Communicate and Move Rapidly?

- Some plants release volatile chemicals when attacked. These chemical attract insect predators that can remove the threat (**Figure 44-15**).
- Injured plants release volatile chemicals that signal neighboring plants to produce substances (e.g., salicylic acid) that protect them from predation or infection.
- Some plants, such as the Venus flytrap and bladderwort, can move rapidly to capture prey. In touch-sensitive plants such as the Venus flytrap, movement occurs when touch sensors that stimulate ion flow are triggered, resulting in the rapid loss of water in specialized cells that cause petiole movement (**Figure 44-16**). Bladderworts generate lower pressure in their bladders that suck in prey when bristles around the bladder openings are triggered (**Figure 44-17**).

LEARNING OBJECTIVES

Section 44.1 What Are Plant Hormones, and How Do They Act?

- Explain the difference among phototropism, gravitropism, and thigmotropism.
- Discuss the factors influencing the action of plant hormones.
- Describe the functions of auxins, gibberellins, cytokinins, ethelene, and abscisic acid.

Section 44.2 How Do Hormones Regulate the Plant Life Cycle?

- Discuss the interplay between abscisic acid and the environment resulting in the germination of seeds.
- Describe the role of auxin in the gravitropism of seedling development.
- Discuss the role of auxin in root growth.
- Discuss the interaction between auxin and cytokinin in determining the amount of stem and root branching.
- Describe the flowering of day-neutral, long-day, and short-day plants.
- Describe how plants detect light.
- Discuss the various roles of phytochrome.
- Discuss the role of ethylene in fruit ripening.
- Describe the interaction of cytokinin, auxin, and ethylene production in the initiation of plant dormancy.

Section 44.3 Can Plants Communicate and Move Rapidly?

- Discuss how plants can call in insects to help deter pests.
- Discuss the concept of plants having an immune system.

QUIZ 1

1. The downward growth of roots into the ground is an example of _____.
   a. phototropism
   b. gravitropism
   c. abscission
   d. senescence

2. Which hormone is responsible for stimulation of cell division and cell differentiation as lateral buds are released from dormancy?
   a. cytokinin
   b. auxin
   c. ethylene
   d. abscisic acid

3. Long-day plants flower specifically when the _____.
   a. light period is less than some critical length
   b. light period is greater than some critical length
   c. dark period is less than some critical length
   d. dark period is greater than some critical length

4. Gravitropism in roots is a response to _____ and to the hormone_____.
   a. gravity, auxin
   b. Earth's rotation, cytokinin
   c. altitude, ethylene
   d. the biological clock, phytochrome P_r

5. Rapid stem elongation and larger fruit size are promoted by _____.
   a. auxins
   b. gibberellins

   c. cytokinins
   d. abscisic acid

6. Phytochromes are believed to be play a role in _____.
   a. inhibition of the elongation of seedlings in the case of P_{fr}
   b. detection of light wavelengths that control the metabolic "biological" clock of plants
   c. chlorophyll synthesis
   d. all of the above

7. Within a plant cell that has statoliths concentrated at the bottom of the cell, the highest concentration of auxin would be found
   a. at the upper end of the cell.
   b. in the middle of the cell.
   c. at the bottom of the cell.
   d. evenly distributed throughout the cell.

8. A shoot's apex effect of keeping the lateral buds dormant is termed apical
   a. dormancy.
   b. dominance.
   c. suppression.
   d. superiority.

9. In the cycle of photoperiodism, far-red light is absorbed
   a. by both forms of phytochrome.
   b. by neither form of phytochrome.
   c. only by the red form of phytochrome.
   d. only by the far-red form of phytochrome.

10. When the phytochrome in the red form is exposed to red light, it will
    a. be converted into the phytochrome in the far-red form.
    b. be broken down into a physiologically inactive compound.
    c. stay in the phytochrome red form.
    d. reflect the red light.

11. Flowering in a long-day plant can be produced by an uninterrupted light period of
    a. 12 hours and an uninterrupted dark period of 12 hours.
    b. 16 hours and an uninterrupted dark period of 8 hours.
    c. 8 hours and an uninterrupted dark period of 16 hours.
    d. all of the above.
    e. none of the above.

12. Flowering in a day-neutral plant can be produced by an uninterrupted light period of
    a. 12 hours and an uninterrupted dark period of 12 hours.
    b. 16 hours and an uninterrupted dark period of 8 hours.
    c. 8 hours and an uninterrupted dark period of 16 hours.
    d. all of the above.

13. Which of the following light cycles could result in a long-day plant NOT flowering?
    (LP = light period, DP = dark period and FOR = flash of red light, FOFR = flash of far red light and FOW = flash of white light) These occur in the chronological order specified in each response. All times are in hours.
    a. LP = 8, DP = 6,
       FOR, DP = 5, FOR, DP = 5
    b. LP = 8, DP = 6,
       FOW, DP = 5, FOW, DP = 5

c. LP = 8, DP = 6,
   FOFR, DP = 5, FOFR, DP = 5
d. all of the above
e. none of the above

14. Which type of signal is responsible for the rapid movement of plant leaves of the mimosa or Venus flytrap?
    a. electrical
    b. auxin
    c. cytokinin
    d. ethylene

15. When the Venus flytrap snaps shut around a small insect, what is the response of the plant to the initial electrical fluctuation?
    a. Electrical potential from the hair movement stimulates cells of the outer layer to pump hydrogen ions rapidly into the cell walls. This creates an acidic condition that activates enzymes that loosen fibers in the cell walls, weakening them. High osmotic pressure inside the cells causes them to absorb water and rapidly increase in size while the inner layer does not. This snaps the leaf closed.
    b. Motor cells increase their permeability to potassium ions, water flows by osmosis, and the cell shrinks as water is lost from the leaflets and petiole and they wilt rapidly.
    c. Potassium and hydrogen ions are pumped across the cell membrane, resulting in a rapid increase in water from osmosis. The leaf snaps shut with the increase in turgor pressure.
    d. The insect is held in place by sticky substances until the leaf can respond to the electrical signal in a manner similar to an animal nerve cell.

QUIZ 2

1. One rotten apple spoils the whole barrel because the rotting apple produces _____.
   a. auxin
   b. cytokinin
   c. gibberellin
   d. ethylene

2. One of the possible consequences for a plant with a defective gene for the production of abscisic acid would be _____.
   a. an increased rate of senescence of the plant's tissues
   b. an increased rate of lateral root development
   c. the development of fruit without fertilization
   d. a decreased dormancy period for the plant's seeds

3. If the tip of a plant shoot were removed, the lateral buds near the top of the shoot will
   a. die from the lack of an important resource.
   b. remain dormant.
   c. begin to grow into branches or flowering stalks.
   d. develop into roots instead of stem branches.

4. In experiments with plant growth, scientists have found that a plant tip will bend toward the light because
   a. when the shoot tip is not exposed to light, it will not produce auxin.
   b. auxin will move to the darker side of the stem and cause more cell elongation there.

c. auxin will move to the lighter side of the stem and cause less cell elongation there.

d. the cap provides the necessary support to hold up the plant.

5. If the source of cytokinin were eliminated during the development of a plant, the lateral buds would
   a. remain dormant.
   b. produce branches.
   c. produce roots instead of branches.
   d. produce flowering stalks instead of branches.

6. Photoperiodism is a response to
   a. light and dark periods.
   b. humidity.
   c. temperature.
   d. carbon dioxide levels.

7. Why do the coleoptiles of oats bend toward the light?
   a. Oat coleoptiles exhibit negative phototropism.
   b. Gibberellins stimulate growth on the illuminated side of the shoot.
   c. Auxins are found at highest concentrations and stimulate growth on the darker side of the shoot.
   d. Changes in turgor pressure in cells on the darker side of the shoot cause the shoot to bend toward the light.

8. If a plant were strapped upside down to the wall of a rapidly rotating drum (like the ones you might ride on at an amusement park), which of the following would you expect to happen?
   a. The artificial G forces would cause the roots to grow toward the middle of the drum and the stem to grow away from the middle.
   b. The artificial G forces would cause the roots to grow away from the middle of the drum and the stem to grow toward the middle.
   c. The roots and stem would curl around such that the roots would grow toward Earth and the stem would grow toward the sky.
   d. The roots and stem would grow in the direction they were oriented because they could no longer be influenced by Earth's gravitational forces.

9. Which of the five classes of plant hormones is MOST likely to play a prominent role in the maintenance of dormancy?
   a. auxins
   b. gibberellins
   c. cytokinins
   d. ethylene
   e. abscisic acid

10. How does auxin control the direction of root growth so that roots grow down into the ground, rather than up out of the ground?
    a. Roots respond to light penetrating the soil in a manner known as *positive phototropism*.
    b. Auxin accumulates along the side of the growing root farthest from the ground surface. This

inhibits cell elongation on the lower side of the root, and the root bends and grows downward.
    c. Auxin stimulates a negative gravitropism.
    d. All of the above are correct

11. What is believed to be responsible for the detection of gravity in roots and shoots?
    a. Plastids settling onto the lowest surface of the root/shoot (i.e., farthest away from the surface of the soil) initiate a series of steps that end with auxin triggering an unequal elongation in the root/shoot.
    b. Auxin itself is able to detect the direction of gravity.
    c. All of the above are correct.
    d. None of the above is correct.

12. What is responsible for ensuring that the relative growth rate of the roots does not overwhelm the growth rate of the shoots and vice versa?
    a. auxin
    b. cytokinin
    c. apical dominance
    d. (a) and (b)

13. It is NOT true that gibberellins
    a. promote the development of fruit.
    b. activate enzymes that stimulate seed germination.
    c. stimulate stem elongation.
    d. delay the process of senescence (aging).

14. Although different from animal antibody-producing immune systems, plants produce many chemicals that aid in protection from insects, herbivores, and diseases.
    a. True
    b. False

15. Which of the following may happen to plants that are downwind from a plant of a different species that is under insect attack?
    a. Nothing will happen, since they are two different species of plants.
    b. The downwind plants may respond to some of the volatile signal compounds produced by the plant under attack, increasing production of protective compounds before being attacked.
    c. The plant undergoing attack will produce additional auxins and replace the tissues under attack, but the downwind plants will not be affected.
    d. The plants downwind from the plant under attack will most likely be swarmed with insects before any compounds can be produced by the plant under attack.

ANSWER KEY

Quiz 1

1. b
2. a
3. c
4. a
5. b
6. d
7. c
8. b

9. d
10. a
11. b
12. d
13. c
14. a
15. a

Quiz 2

1. d
2. d
3. c
4. b
5. a
6. a
7. c
8. b

9. e
10. b
11. a
12. d
13. d
14. a
15. b

NOTES

NOTES

NOTES

NOTES

NOTES

NOTES

NOTES

NOTES

NOTES

NOTES